The Anatomy of Xerography

The Anatomy of Xerography

Its Invention and Evolution

J. Mort

McFarland & Company, Inc., Publishers
Jefferson, North Carolina, and London

British Library Cataloguing-in-publication data are available

Library of Congress Cataloguing-in-Publication Data

Mort, J.
The anatomy of xerography : its invention and evolution / J. Mort.
p. cm.
[Includes index.]
Includes bibliographical references.
ISBN 0-89950-442-6 (lib. bdg.: 50# alk. paper) ∞
1. Xerography. I. Title.
TR1045.M67 1989
686.4'4—dc20 89-42739
CIP

Printed in the United States of America.

McFarland & Company, Inc., Publishers
Box 611, Jefferson, North Carolina 28640

"While from the bounded level of our mind,
Short views we take nor see the lengths behind"
Alexander Pope, *Essays in Criticism*

Dedicated to everyone
who contributed through the ages—
but especially Chester Carlson

Table of Contents

Preface

In the fifty years since the first xerographic copy, "10.-22.-38 Astoria", was made by Chester Carlson and his assistant, Otto Kornei, in Astoria, New York, xerography has established itself as one of the major technological innovations in the history of man. Today billions of dollars in revenues and several trillion documents are generated annually by this technology. The technology has evolved and diffused into society to such an extent as to transform the means of disseminating and, more recently, creating printed information. Millions are already totally familiar with using the ubiquitous copier and are becoming so with that growing adjunct to word processors, the laser printer. Less familiar perhaps, is the broad range of technical factors involved in the functioning of this technology.

The major objective of this book is to inform the reader about the science and technology that led to Carlson's invention. Consequently this is not primarily a description of the phenomenal business success of xerography or the Xerox Corporation, although they are inevitably and rightfully part of the story since business and market considerations are as crucial as technology for any successful innovation. It is my belief that it is possible and important for both the technical expert and less technically inclined, especially those who must make decisions in an increasingly technological context, to have a better understanding of science and technology and particularly their role in innovation. Accordingly I have chosen to discuss the invention, innovation and evolution of xerography from this perspective. I have not shrunk from discussing the sophisticated technical concepts essential to appreciate the technology and continued contributions of research, together with their relevance to the needs of the marketplace. I have tried throughout to articulate these matters in as descriptive a way as possible and the experts will recognize where the goal of clarity rules over precision, but hopefully not accuracy.

An ancillary purpose is to use this account to illustrate how "high tech," as it is colloquially called, happens. It is neither purely a case of "eureka" nor of unfathomable mystery. Rather it is the coordination of creativity, collective understandings and sometimes luck with an ability to conceptualize and orchestrate diverse technical capabilities to meet real or perceived needs. Typically though, in the final analysis one individual, through a unique creative insight coupled with an unshakable conviction in his or her idea, is the synthesizing spark – Chester Carlson was such an individual.

I should like to express my appreciation to several of my colleagues, Dr. Inan Chen, Dr. Dan Hays, Dr. Eric Pell, Dr. Alan Perregaux, Dr. Paul Roetling, Dr. Michael Shahin and Arthur Zuckerman at the Xerox Corporation who were kind enough to read various parts of this book. Their helpful comments and suggestions materially helped with respect to both its clarity and accuracy. Any errors or inadequacies that remain are mine alone. Equally, I was fortunate to have the assistance of Rick Pieramico and Don Cobb of the Art and Photographic Services of the Xerox Webster Research Center.

Although a review was required by both the legal and public relations departments of the Xerox Corporation before the manuscript could be submitted to a publisher, the presentation of facts or opinions expressed in this book are those of the author and should not be construed to represent the views, strategic intent or business plans of the Xerox Corporation.

J. Mort,
Webster, New York
August, 1988

Introduction

On October 22, 1938, in Astoria, New York, Chester Carlson produced the first copy of an original document by a new marking technology, to be called xerography. During that same decade, scientific research was underway which would ultimately lead to solid-state microelectronics. The subsequent history of these two developments constitutes a fascinating example of the vagaries which can and often do occur in the interplay of science, technology and innovation. Thus, although the practical feasibility of xerography, demonstrated in 1938, preceded the invention of the transistor by about a decade, in actual commercialization the positions were reversed. Further the basic understanding of the materials and phenomena of solid-state electronics far outstripped those important to xerography, despite the fact that the modern theory of solids is central to both technologies.

The disparate progress in these technologies had several origins, but none as important as the nature of innovation. Innovation in this context means the profitable marketing of a product based on materials or phenomena, or a combination of both. Solid-state electronics has been identified as a rare example of an innovation based on science.[1] Insofar as one can dissect the complex, often symbiotic, relationship between science and technology, xerography is an example at the other extreme. A more important determinant of a successful innovation than its genesis, however, is that it must either satisfy a definable or perceivable need, or create a demand.[2] The phenomenal commercial success of solid-state electronics and xerography came about despite the fact that, although in both cases a need or market-pull was identified, the full magnitude and richness of their market potential were not initially appreciated. Xerography and its innovation were extraordinary in the degree to which this discrepancy existed.

Another reason for the different rates of development in these two technologies had to do with the imperatives of the customers' needs. In the case of solid-state electronics, there was a large and existing market for electronic rectifiers, amplifiers and switches based on thermionic vacuum tubes. These had major shortcomings in terms of size, dependability and energy consumption. Moreover a major and growing customer was the military with substantial funding resources at its command. By contrast the need for improved or more efficient ways of making copies of documents was originally perceived, except perhaps by Carlson, as more of a convenience than a necessity. Carbon copies were sufficiently legible, if not aesthetically pleasing, and appeared to fulfill the general need for cheap, convenient copies. Thus there was no obvious major market opportunity recognized or stimulating force to create one except, as events proved, the invention itself. For these and other reasons, the two innovations developed along different paths and for distinct purposes. Today, fifty years later, their paths have converged to synergistically fuel a new innovation which has been called document processing. For no one can doubt that with the explosive growth in the processing, transmission and distribution of electronic information this is the Age of Information.[3,4] This revolution has been brought about by the computer and particularly its evolution in terms of capability, coupled with size and cost reductions. The basis of this technological innovation, whose influence pervades our everyday life, is solid-state electronics in general and the transistor in particular, both of which are enabled by crystals of silicon. Modern silicon processing techniques which allow the efficient mass production of integrated solid-state devices of previously unimagined sophistication have enabled both the miniaturization and low cost of today's desktop computers. The creation of information now involves electronic circuits executing millions of commands per second. On the input end, this involves the use of keystrokes to produce digital data, or the optical scanning of documents to convert written information into electronic bit streams. Ultimately, this digital information has to be interfaced with humans whose data acquisition and processing are much slower. At the output end, therefore, the information needs to be transformed into a visual image such as temporally on a video display terminal or VDT. This may be a cathode ray tube (CRT) or a liquid crystal-based display. However, the cheapest, most versatile and dominant display medium for information flow, now and for the foreseeable future, is a permanent document created by making marks on paper.

It is the intent to describe the emergence of the Information Age primarily from the perspective of xerography, its technical underpinnings and evolution from a means of copying documents to the emerging and growing phase of their creation and dissemination. It will become apparent that this technology utilizes a far wider range of phenomena than semiconductor electronics whose developments will not be extensively discussed, but will serve as a significant counterpoint to the main theme. Since the emergence of xerography, other marking technologies such as ink-jet, direction deposition and thermal transfer processes have been developed and commercialized, but no in-depth comparative discussion will be given.

Although in certain products, particularly those of lower cost, these technologies have had some success, none can match the broad applicability of xerography. After almost forty years of commercial application, xerography has demonstrated an impressive ability to maintain the balance between cost and quality while being increased over a factor of ten in speed.

Compared to solid-state electronics, xerography involves significantly different phenomena, materials' properties and device fabrication. It employs a large optical to electrical transducer called the photoreceptor which typically has an area of one or more sheets of paper. This required the development of radically new and different materials, namely, amorphous solids. Initially this was done empirically and, in contrast to electronic semiconductors, it was the technology that gave impetus to the birth of a science—that of the disordered or amorphous state. This class of solids, probably first produced by man inadvertently in prehistoric times with a fire on a beach, only became the subject of intensive scientific study in the 1960s. This, despite the fact that many of the questions regarding the electrical and photoelectrical properties of disordered materials are the same as for crystalline semiconductors.

In addition, a number of hitherto ignored or neglected areas of scientific research were regenerated by Carlson's invention. These included electrostatics, the physics of coronas, and triboelectricity. Moreover the various elements of this technology originated both in the mists of antiquity and the most modern scientific concepts. Thus it involves the frictional electrification of solids, well-known to the Greeks, and the chemical, physical, and photoelectronic properties of amorphous solids. Equally important were the emergence and growth in the science and technology of organic materials, particularly polymers, and more recently the transition of that miracle of light, the laser, from a laboratory instrument to a more or less commodity device.

It is important to understand the technical environment in which Carlson found himself in the 1930's, since the invention of xerography did not appear out of thin air. As is characteristic of most inventions, it was stimulated by the perceived shortcomings of the existing technologies; specifically in this case those for making copies. The elements of the solutions were found in an array of apparently unrelated phenomena. These had involved investigations, conducted over centuries, which ranged from practical trial and error to the most creative and sophisticated scientific research. To understand these phenomena and appreciate the significant role they played in the invention, it is necessary to trace a number of complicated threads from the earliest days of science to the 1930s. This serves to emphasize the dependence of most advances on prior discoveries and the particular diversity of the phenomena involved in xerography. The focus will be on the essential conceptual advances, since it is obviously impossible to pursue, ab initio, all the multitude of developments actually involved.

Chapter 1
Early History: B.C., Before Carlson

A little reflection will establish that all human endeavor is essentially an incremental activity and that progress occurs at the margins. Inevitably, this means that advances occur from a previously accumulated body of insights. Science and technology are no exception to this rule; as Newton said, "I have stood on the shoulders of giants." This is not to imply that progress occurs in obvious or even predictable ways, since history abounds with examples of the seminal ideas generated by geniuses such as Newton, Einstein, Bohr and others. It does mean that the issues, problems and solutions which scientists and technologists grapple with are conditioned by the established data base of knowledge. If any doubt exists about this, it is worth reflecting on the fact that the average person of today knows far more in general scientific terms than the giants mentioned above even dreamed about. Thus concepts such as the double helix, silicon chips, microwaves, liquid crystals, vitamins, etc., are part of the vernacular.

The roles that science and technology play in innovation are the subject of much confusion, since the terms are used in ways to suggest that these are divisible activities or even that science is an inevitable precursor to technology. As a result attempts are made to classify, somewhat arbitrarily, activities such as "basic or pure" science, applied science, and technology. These distinctions, which often appear to loom so large in the physical and chemical sciences, are much less obvious in other fields such as medicine, where research results inevitably bear on the technology of caring for the sick. While such classifications may have some value, the impression given is that there is a fundamental difference in the nature of the methodologies involved. In truth, the distinctions are more to do with motivation than style. Science has to do with the acquisition of knowledge or understanding as an end in itself. Technology is the acquisition and application of knowledge for the purpose of producing a practical product. The methods may be the same but the goals are different.[5,6] As already suggested, science can be a catalyst for an innovation, as was the case for solid-state electronics. It would be quite wrong to presume that this is the normal flow of events. More commonly, purely empirical approaches can lead to

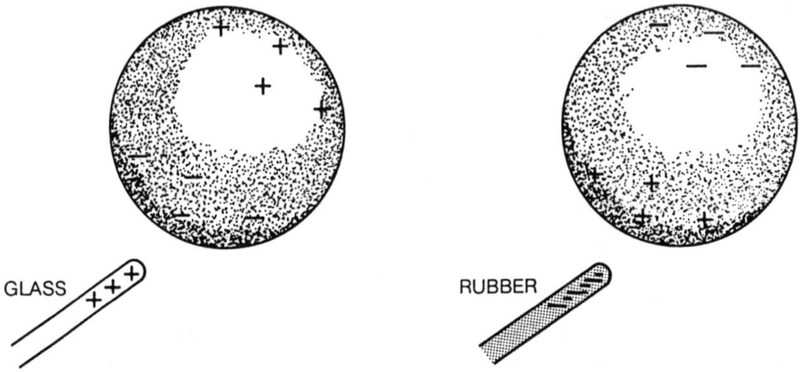
GLASS
RUBBER

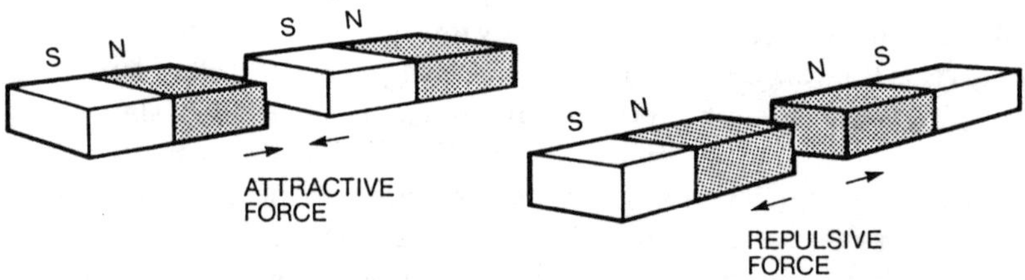
S N
S N
ATTRACTIVE
FORCE
S N
N S
REPULSIVE
FORCE

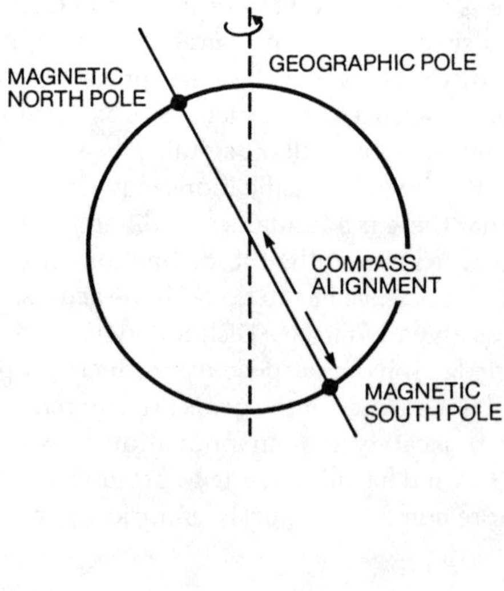
GEOGRAPHIC POLE
MAGNETIC
NORTH POLE
COMPASS
ALIGNMENT
MAGNETIC
SOUTH POLE

an innovation in the absence of any clear understanding of the basic principles involved.[7,8] Once an innovation is successful, however, the empirical try and test approach can become very inefficient and costly, particularly if one is challenged by competition. In such a situation, the establishment of a body of understanding that can lead to quantitative specification of the critical parameters for materials, processes, device operations or market forces is vital.[9] This certainly proved to be the case for xerography.

Hundreds of years before Christ, the Greeks observed triboelectricity, the frictional production of static or stationary electric charge, in that a fossilized bituminous resin called amber when rubbed with silk was seen to attract small pieces of straw or hair. Anyone who has been jolted by the self-generated charge produced by walking over a nylon rug or frustrated by the static cling of clothes taken from a drying machine will recognize the phenomenon. It was also observations with a naturally occurring iron oxide ore, the lodestone, which revealed the phenomenon of magnetism and over the centuries the ability of lodestone to induce magnetic properties in certain metals was discovered. If such a magnetized rod was suspended freely on a string, it was observed to line up in a certain direction with respect to the earth and thus the compass was invented. Depending on their orientation with respect to each other, two such magnetized rods either attract or repel each other. Thus the north pole of a magnet which in alignment points to the geographic north repels a like north pole but attracts a south pole. For a long time electrostatic attraction was often confused with that of magnetism and the differences between the two were not experimentally established until the mid-sixteenth century.[10]

In 1600, William Gilbert, physician to Queen Elizabeth I, showed that many other bodies besides amber could be made to attract light substances if rubbed vigorously, particularly in very dry weather. He also invented an instrument called the versorium which consisted of a thin metallic needle pivoted on a support, as in a compass. On bringing up a rubbed solid, the attracted needle would swing toward it. Such an instrument and its sensitivity greatly improved the detection and measurement of the effect and allowed Gilbert to identify the class of materials which had these properties in common with amber. Following his use of the Latin word for amber, *electrum*, these were called electrica (electrics), whereas those that did not have amber's attractive properties were termed anelectrica

Opposite: *Glass rubbed with silk or rubber rubbed with fur acquires positive and negative charge respectively which can induce counter-sign charges in the solid spheres. Simple bar magnets also attract or repel depending on the relative alignment of their north and south poles so that a magnetic compass needle aligns with the north and south magnetic poles of the earth. Such similarity in behavior between electrostatics and magnetism, which play a critical role in xerography, led to confusion between these two distinct natural phenomena until the 16th century.*

(nonelectrics). He also noted that such charged bodies could even attract the smoke from an extinguished candle. Three and a half centuries later this seemingly trivial observation would be exploited in xerography. Gilbert also conducted extensive investigations on magnetism which he published in *De Magnete* and for which he is most remembered. Amongst his findings was the fact that the earth itself is a magnet, which accounted for the behavior of the compass.[10] In the context of the present discussion, modern day xerographic copiers critically depend on the combined use of electrostatic and magnetic phenomena.

Throughout the seventeenth and eighteenth centuries, other observations were made relating to the properties of static electric charge, as opposed to electric current which is the motion of such charge. The first tentative connection between these two forms of electricity was made with the invention in 1660 by von Guericke of a frictional engine to produce a constant supply of electricity. In 1729, Gray took a crucial step forward by distinguishing between good electrical conductors (nonelectrics) and poor conductors (electrics). Until his research, essentially no experiments had been carried out concerning whether charge, or "electrick vertue" as it was known at the time, could move *through* a material. In a series of simple yet elegant studies, Gray showed that electrick vertue could be transmitted over at least 100 feet in substances such as copper wire. By contrast, no such motion was observed with glass, silk and like materials. Today, nonelectrics and electrics are known as conductors and insulators, respectively. Thus, for the first time a differentiation of solids on the basis of their ability to conduct electricity was made.[10]

Shortly afterwards it was noticed that charged bodies may repel as well as attract. This symmetry of electric force was the first indication that two types of charge might exist, although it was not until about 1730 that the French scientist Du Fay established that the charge induced by rubbing could be of two types. Much of our understanding and nomenclature of electrostatics is due to major contributions of Benjamin Franklin, for which efforts he was elected a Fellow of the Royal Society. Thus it was Franklin who coined the terms positive and negative for the two types of static electricity. In addition, his much more celebrated kite-flying experiments in thunderstorms, an exceedingly hazardous undertaking as others found to their cost, established that lightning is a gargantuan transient electric current produced, as we now know, by the discharge of static electricity frictionally built up within storm clouds. More pertinent to the phenomenon of triboelectricity was his realization that the sign of the frictional charge produced depended on the nature of *both* substances rubbed together.[11]

Around 1785, Coulomb explored the relationship determining the magnitude of the force between two electrostatic charges. Coulomb's law which encapsulates the results of these experiments states that the force is

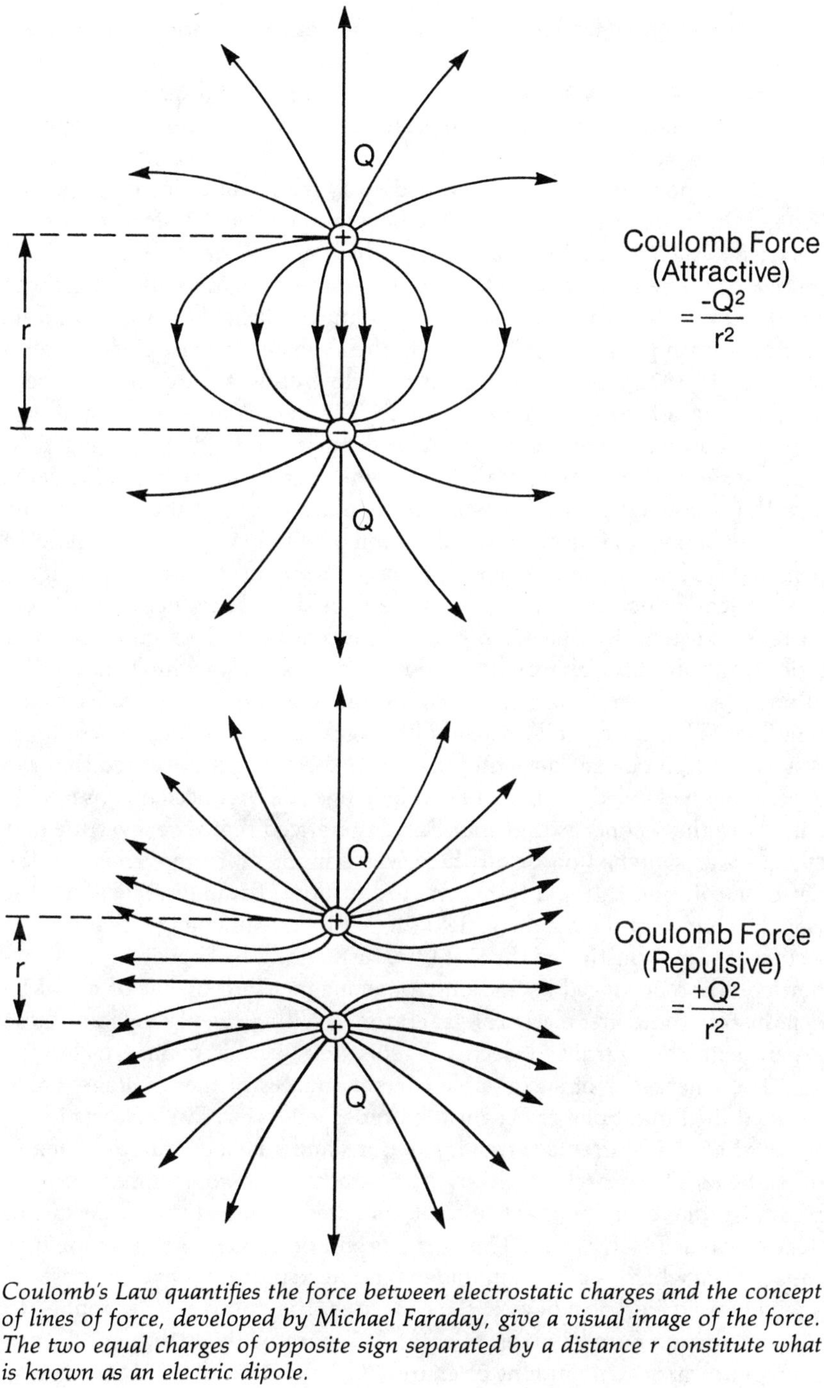

Coulomb's Law quantifies the force between electrostatic charges and the concept of lines of force, developed by Michael Faraday, give a visual image of the force. The two equal charges of opposite sign separated by a distance r constitute what is known as an electric dipole.

proportional to the product of the magnitude of each charge and inversely proportional to the square of the distance between them. Thus the so-called coulomb force increases as their separation decreases and is repulsive or attractive depending on whether the charges have the same or opposite polarity, respectively.[12,13]

At this point the investigations led away from static electricity per se. Instead the emphasis centered on the development of chemical sources of electricity, such as the Volta batteries, which arrived on the scene around the end of the eighteenth century. This led to the recognition that there was a connection between electric currents and magnetic fields.[12,13] This important relationship was first discussed by the Danish physicist Oersted, who observed in 1820 that an electric current flowing in a wire could cause a deflection in a nearby compass needle.[10] The inescapable conclusion was that the current produced an associated magnetic field. Following this, Ampere showed that two parallel wires with currents flowing in the same direction attracted each other while repulsion occurred if the two currents flowed in opposition. In other words, a coil of wire in which a current flows behaves as a bar magnet. Michael Faraday was the first to conceive, in 1831, of electrical induction whereby a magnetic field might induce an electrical current indistinguishable from that produced by Volta's chemical batteries.[14] Faraday took an iron ring and wound two coils of wire diametrically opposite each other. One coil was connected to a battery to produce a current flow. The second coil, instead of being connected to a current source, was wired to a current detecting instrument. At first it appeared that the experiment had failed. A current flowing in one coil produced no observable current in the second, except that Faraday noticed that whenever the first current was switched on or off there was a momentary movement in the detector coil, indicating a transient current flow. Faraday realized that it was the change of the magnetic field, in fact its rate of change, that creates a current. This constituted the first transformer and by showing that a current could be produced by moving a bar magnet into or out of a coil he demonstrated the first electrical generator.[14] Within twenty years of these experiments, decentralized electrical generators were in commercial use.[10]

The emergence of sustainable current sources of high voltage which avoided the limitations of chemical batteries led to increased efforts in the study of electric currents through solutions and gases. Again, as in many other fields of research in electricity, Faraday, the consummate experimentalist, played a dominant role.[14] He studied what he termed specific inductive capacity of matter. This is the origin of a phenomenon known as capacitance which is central in understanding xerography and the basis of a widely used electrical device, the capacitor. If a voltage, V, is applied to two conducting parallel plates of area, A, separated by an air gap, d, then each plate carries an amount of charge, Q, such that $Q=CV$ where C is

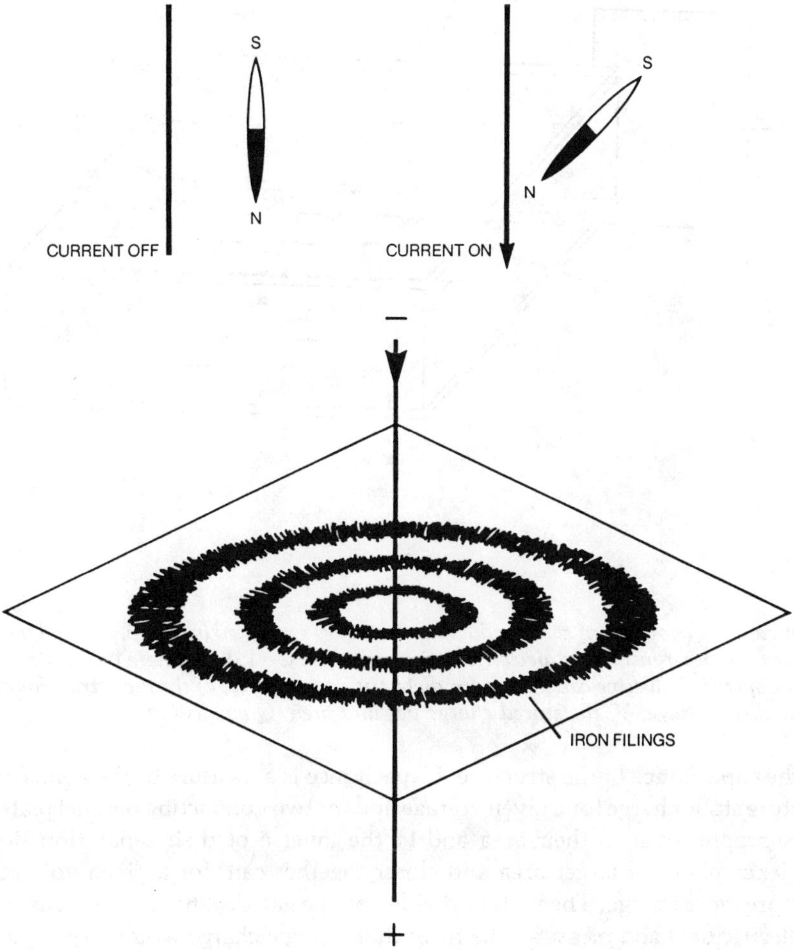

A flow of electric charge or current produces a related magnetic field as detected by the magnetic needle. Ferromagnetic materials such as iron filings align themselves along the associated latent magnetic lines of force, thus making them visible.

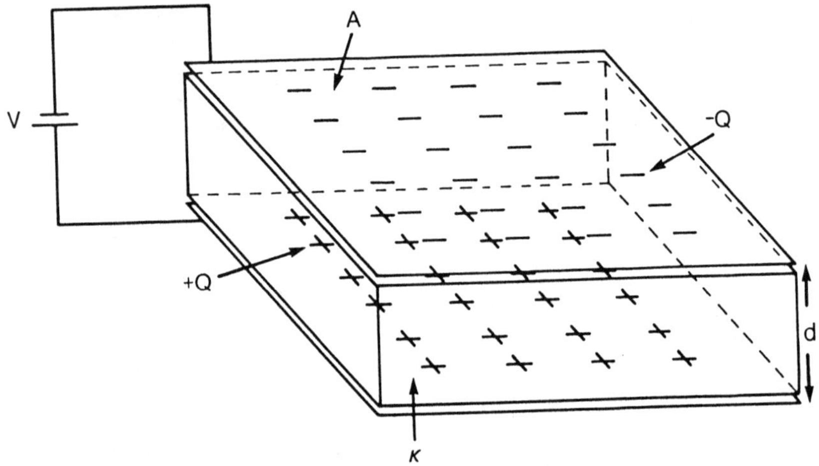

A capacitor consists of two conducting electrodes separated by a gap, d, filled with air or another insulating dielectric. Its capacitance, $C=kA/d$, where k is related to the specific inductive capacitance or dielectric constant, K, *of the dielectric. For an applied voltage, V, the stored charge per unit area, Q equals CV.*

the capacitance of the structure. Capacitance is a measure of the ability to store static charge for a given voltage and for two conducting parallel plates is proportional to their area and to the inverse of their separation.[12,13] Hence plates of larger area and closer together can, for a given voltage, store more charge. The voltage divided by the gap distance is known as the electric field and measures the force that a single charge would experience if placed between these plates. If an insulating material of thickness, d, is inserted into the air gap the situation changes. For the same applied voltage, the charge carried by each plate increases by a factor, the specific inductive capacitance of the material; today the term dielectric constant, K, is used.[12] Typical dielectric constants range from 2 to 10 for most insulating solids, although a few materials have values in the thousands. If the surface charge rather than voltage is kept constant then the dielectric constant of the inserted material results in a reduced voltage and hence electric field within the dielectric. Faraday also investigated current flow through water or solutions of salts or acids, enunciating the laws of the associated chemical decomposition or electrolysis and began the study of gas discharge at low pressure. He discovered and termed the name glow discharge for the luminous zones observed, together with current flow, if

Discharge tubes consist of partially evacuated glass envelopes with metallic electrodes. A voltage applied between the two electrodes produces a discharge, current flow and a glow the color of which is characteristic of the gas used. In addition to the vital role that such relatively simple devices played in revealing the structure of atoms, they were a source of much entertainment for the well-to-do in the 18th century.

a voltage of several hundred volts is applied between electrodes in tubes filled with air at a fraction of atmospheric pressure.[14] Today's fluorescent lamps are a practical application of this phenomenon. It was probably Faraday's close involvement with both electrolysis and gas discharges that led to their accepted synonymity for several decades. Thus the terms anode, cathode, anion and cation were first created to explain electrolysis. Despite this and the clear recognition that electrolysis involved the motion of electrical charge, Faraday did not think of the electrical conduction as involving particles. Rather he thought that it involved stresses or strains, as in the bending or stretching of a rod.[14]

During the nineteenth century the theory of waves became a dominant concept in physics. A wave is the propagation of a disturbance through a medium in some periodic or repetitive form.[12,13] The most familiar to us are waves in the ocean. Certain factors characterize a wave: the amplitude of the disturbance (height of the wave), the wavelength (the distance between two consecutive peaks) and the wave velocity. The product of the wavelength, λ, and frequency, f, equals the wave velocity, V, i.e., $V = f\lambda$. The frequency or number of complete wavelengths passing through a point per second is expressed as cycles per second; in honor of the German physicist Hertz, 1 cycle per second is called 1 Hertz or 1 Hz. In the case of ocean waves, the medium through which the waves move is water. Far from shore, while the wave moves forward, the water itself only moves up and down. The oscillation or displacement produces peaks and

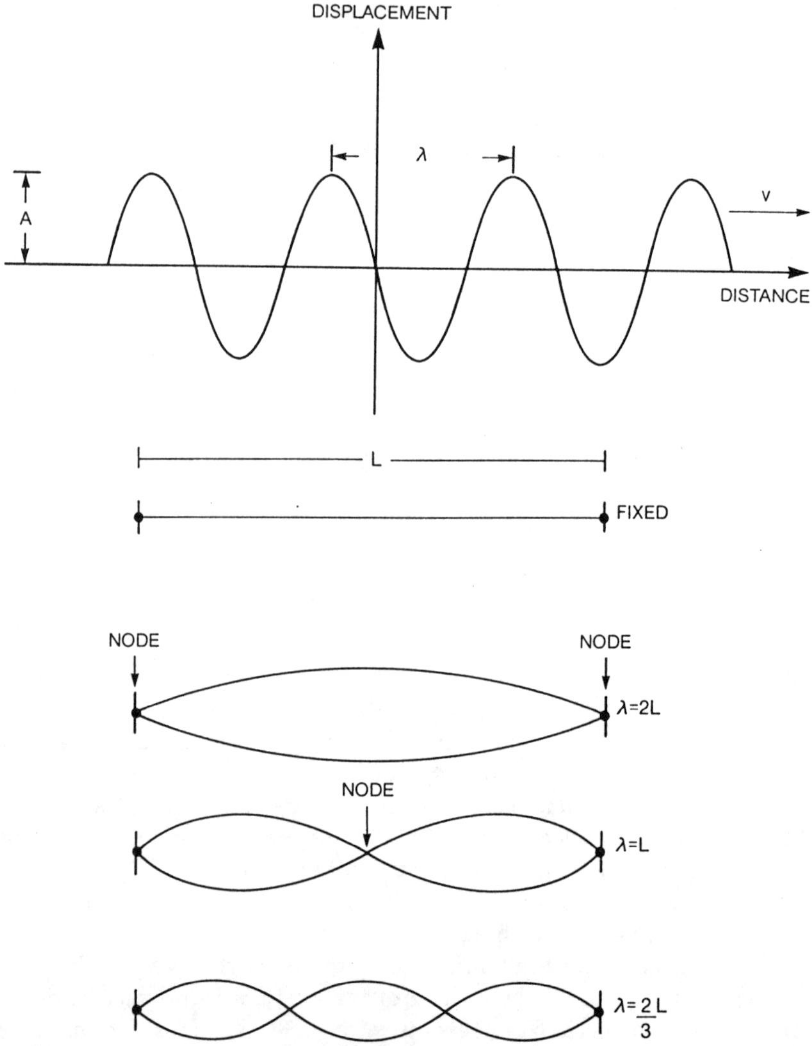

A wave is a periodic disturbance moving through a deformable medium with a velocity v equal to the product of the wavelength λ and frequency f. The amplitude of the wave, A, is the degree of deformation, and A^2 determines in part the energy transmitted by the wave per unit time. A disturbance involving wave motion in the direction of propagation is called a progressive or traveling wave. For other waves involving motion constrained within fixed physical dimensions, the traveling waves are reflected at the boundaries and the initial and reflected waves form standing waves. Stringed instruments are an example where standing waves are important. For a string of length L, a large number of standing waves of different wavelength λ =2L/n, (where n=1, 2, 3 ...) are possible, depending on the energy used to generate the vibration.

troughs in the water about an equilibrium position, which in the absence of the wave, is the flat level of the ocean. From experience playing on a beach, everyone can deduce that the energy of a wave is related to its height (actually proportional to it square). Ocean waves are one example of progressive waves in which forward motion of the wave takes place. Another familiar example of wave propagation, the sound produced by musical instruments, illustrates a second wave-type. Here the medium is, for example, a string and in this type of wave the displacements at each secured end of the string, called nodes, are zero. The frequency or note produced depends on the number of half wavelengths contained within the string length which can be controlled by the position of the fingers. Such waves are said to be standing or stationary waves and have the important characteristic that they do not lose energy unless acted on by an external force. Such an external interaction, as with air molecules, induces corresponding progresive waves in the air which we hear as sound. The fact that we can hear sound, even if the source is out of sight, indicates that sound waves can go around obstacles. On the other hand, experience suggests that light does not behave in this way, since light casts sharp shadows and one cannot normally see around corners. It was experiences with such everyday phenomena that had led Newton to his corpuscular theory of light in which light was conceived as being composed of small, rapidly moving particles or corpuscles.[12,13]

However in 1801 Thomas Young discovered effects with light which dealt such an apparent death blow to the corpuscular theory that for more than a century it was almost totally abandoned.[12,13] Young's experiment consisted of projecting light from a single source (he used sunlight illuminating a pinhole) through two narrow parallel slits onto a screen. If the distance between the two slits was small enough, the image produced on the screen consisted of alternating light and dark lines. This effect could only be explained by thinking of light as a wave. Young's experiments thus established that light does indeed show the properties of waves, if they are sought on the proper scale. The relationship between the frequency and wavelength of light, denoted by ν and λ respectively, is $c = \nu\lambda$ where the velocity of light is usually symbolized by c. Thus in Young's experiment, the bright lines were interpreted as peaks of the waves reinforcing each other (waves in phase), while the dark lines corresponded to the peak of one wave canceling out the trough of the other (waves out of phase). Such a phenomenon is called interference. In fact from Young's experiment, knowing the geometrical distances involved, the wavelength of green light, for example, is found to be about 0.55 microns or 5500 angstroms (1 angstrom or $\text{Å} = 10^{-8}$ cm and 1 micron $= 10^{-4}$ cm or 10^4 Å). By comparison, sound in the mid-audible range has a wavelength of about one foot. This enormous difference in their wavelengths explains why light and sound

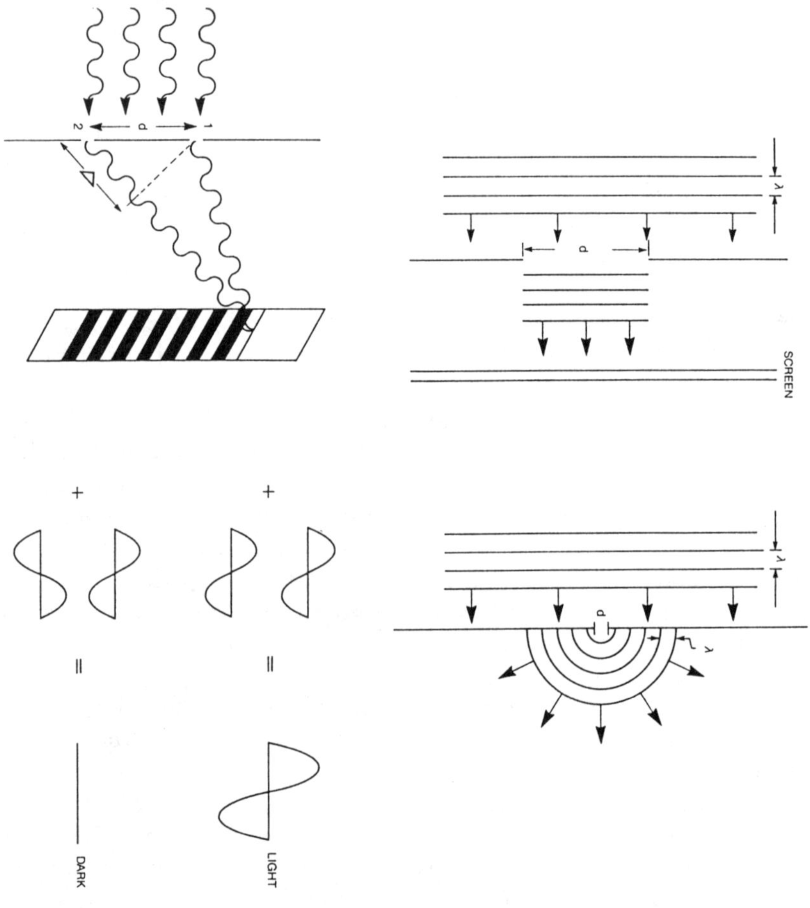

Diffraction or bending of waves is only apparent when an obstacle or aperture has dimensions comparable to the wavelength of the wave. With two or more apertures the waves can cause constructive and destructive interference. Constructive interference, which results in an enhanced disturbance or bright fringe in the case of light, occurs if the difference in the distance traveled by the waves from the apertures to the plane of observation, Δ, is an exact multiple of a wavelength. Destructive interference results in a diminished disturbance or dark fringe if the path difference, Δ, is an odd number of half wavelengths. Since white light contains all colors of the rainbow, each of different wavelength, interference fringes commonly seen with oil on a wet driveway are colored; the oil film thickness floating on the water is comparable or less than wavelengths of visible light. Diffraction phenomena play an important role in certain xerographic laser printers.

appear so different in everyday experience. In 1818 Fresnel showed that if an object placed in a beam of light is made small enough then light can bend or diffract around it. If a large series of such objects are assembled close to each other they form a diffraction grating and produce a series of light and dark lines called a diffraction pattern.

Less clear was the nature of the medium through which light moved since, unlike sound, star light demonstrably moves through the vacuum of space. The question of the medium through which light was propagated was temporarily resolved by invoking the existence of a fictitious medium, the ether, although this was to remain a contentious question.[12–15] In 1864 Clerk Maxwell established the theoretical basis which provided a unifying framework to account for Faraday's observations of electrical induction, his concept of force lines and their relationship to wave motion in general and light in particular.[12,13] It was already clear that there was an intimate connection between electricity and magnetism since a change of an electrical field or a current produces a magnetic field and vice versa. Maxwell showed that charges undergoing acceleration generate an electromagnetic wave which propagates as a transverse wave in which electric and magnetic fields oscillate at right angles to each other and the direction of the wave propagation. Depending on the energy associated with the charge motion, a spectrum of essentially unlimited wavelengths can exist. Such electromagnetic waves, as they are called, were presumed to propagate in vacuum through the ether with the speed of light; a fact which emerged naturally from the theory.[12,13] Experimental verification of Maxwell's theory of electromagnetism was obtained by Hertz in 1888 who detected the electromagnetic radiation from a spark source by using detector coils incorporating their own spark gaps. By adjusting the gap in a detector coil and the latter's position, Hertz was able to observe sparks induced at a distance in the detector coils by the energy radiated from the primary spark and was able to map out the maxima of the electromagnetic waves (brightest induced sparks) by progressively moving the detector away from the source. In this way, he determined the wavelength of what we now call radio waves to be a few feet and established the rudimentary elements of radio communication. Hertz went on to show that the electromagnetic radiation could be refracted by a prism just like light. Since the dimensions of the prism must be much larger than the wavelength of the radiation, the prism was 5 feet in height, made of pitch and weighed 250 pounds![16]

Throughout much of the nineteenth century the prevailing view of electricity was that it involved some form of fluid. This is somewhat surprising since during the same period, dating from 1808, the work of John Dalton had methodically and quantitatively reinvigorated the idea of atoms as indivisible balls, first broached by the Greek, Democritus. In

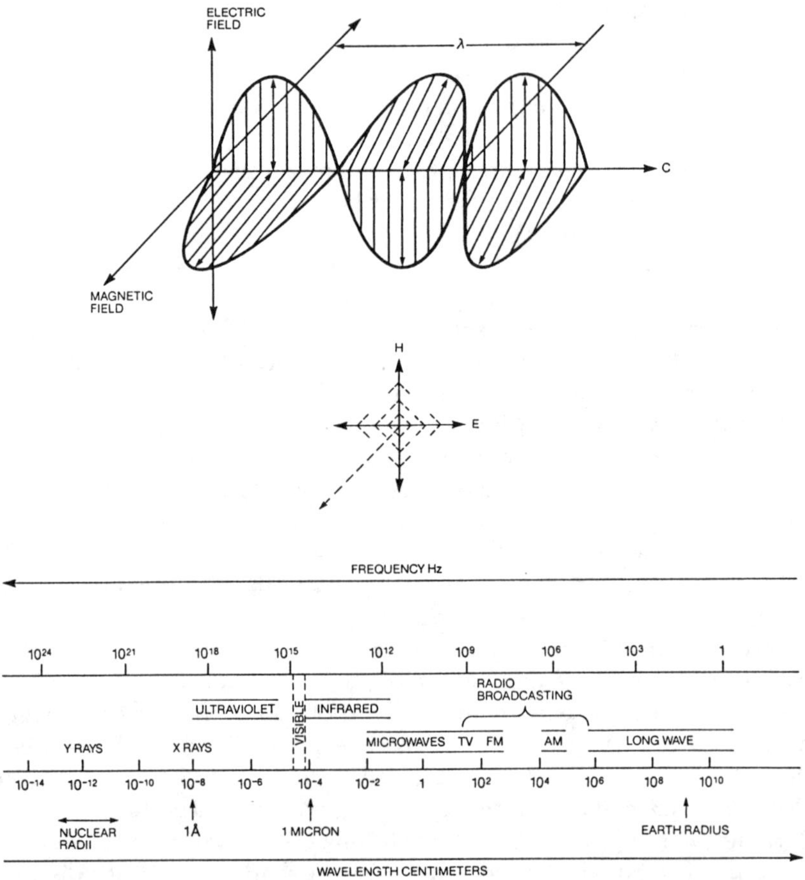

Electromagnetic waves are transverse waves in which electrical and magnetic disturbances occur at right angles to each other and the direction of the wave propagation. The electromagnetic spectrum covers a range of wavelengths, and hence frequencies, of over 10^{24} *(1 followed by 24 zeros). Of this immense range, visible light barely extends over a factor of two.*

Dalton's view, atoms were the smallest part of an element that uniquely identified it and all atoms of a given element would weigh the same but differ in weight from all other elements.[17] Despite this inherently microscopic picture of matter, there was no strong inclination and no experimental basis, to link this chemical view of matter and the physical phenomenon of electricity. This is not to say that efforts were not made to make this connection. Faraday, for one, tried to correlate the ability of metals to conduct electricity with the number of atoms per cubic centimeter. However, he noted that although the element copper has the same number of atoms per cubic centimeter as iron it conducted electricity six times more readily. Faraday also described another paradox in that spherical atoms stacked in a solid, like apples in a box, could not occupy all space and must be embedded in some medium. The problem was that in a metal this medium would have to be conductive whereas in an insulator it would also, implausibly, have to be nonconductive.[14]

The study of gaseous discharges, pioneered by Faraday, was destined to reveal the underlying reality and ultimately provide a unified framework for the understanding of the basis of static electricity, electrical conduction, and the fundamental basis for distinguishing the different classes of solids. In 1858 Plücker observed that some kind of emanation or ray, originating from the cathode and therefore perhaps negatively charged, caused the glow in a gas discharge and produced a fluorescence in the glass wall of the discharge tube. Later the invention of the Geissler and Crookes vacuum tubes and improvements in vacuum pumps led to the observation that these cathode rays could be deflected by electric or magnetic fields in a manner consistent with their being negatively charged. Further the production of shadows on the walls of the discharge tubes by blocking these "rays" and their ability to produce mechanical rotation of paddle wheels began to raise the question as to whether some particle of electricity was involved.[18]

At the close of the nineteenth century, two dramatic discoveries occurred. The first was Röentgen's discovery in 1895 of x-rays which were shown to be electromagnetic radiation, just as light and radio waves. X-rays are therefore of the same nature as light, except they are invisible, of much shorter wavelength and therefore much more energetic. The typical wavelength of light is of the order of 5000 Å while that of x rays is only a few angstroms. The discovery of x-rays was followed closely by Becquerel's observation of the spontaneous radioactive decay of uranium. Both of these mysterious radiations were found to make gases conductive.[19] Within two years of these breakthroughs, J.J. Thomson in Cambridge measured the charge to mass ratio (e/m) for the cathode rays.[20] He found that this value was completely independent of the type of gas in the discharge tube or the metals used as electrodes. Moreover he found the e/m ratio was about 1700 times greater than the charged hydrogen, now

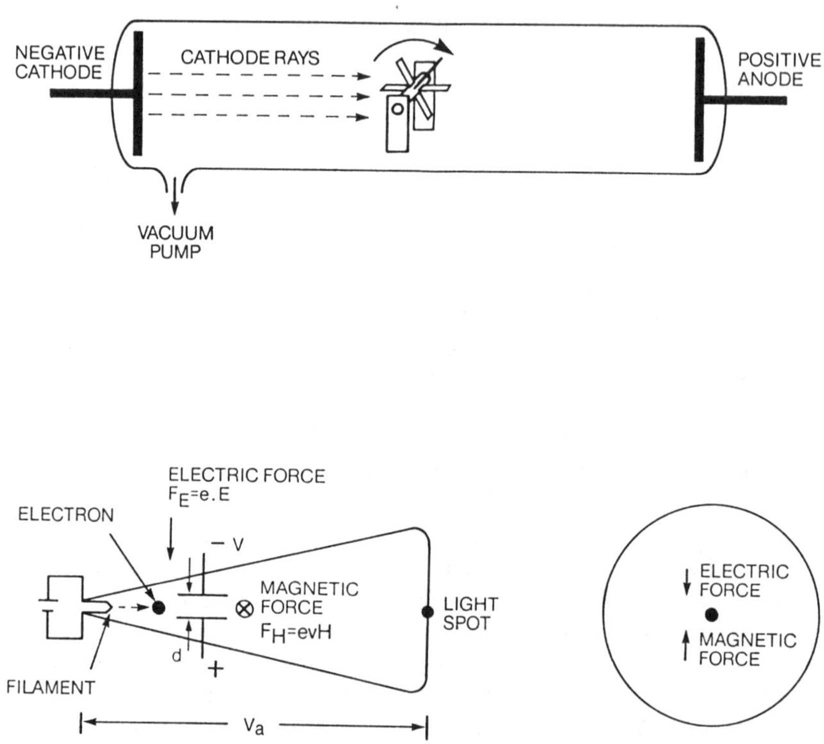

Top: *The glow in a discharge tube was found to be associated with something emanating from the cathode. These so-called cathode rays are deflected by both magnetic and electrical fields in such a direction that they must have an associated negative charge. When these "rays" were found to produce rotation in a paddle wheel, it became apparent that they might really be particles.* Bottom: *The basic outline of the apparatus used by J.J. Thomson in his discovery of the electron. The applied electric field, E, deflects the negatively charged particle downwards with a force eE, (e is the charge on the electron and E=V/d, where V is the voltage applied to the electrodes separated by a distance d). The magnetic field deflects the electron upwards with a force evH, (v is the velocity of the electron controlled by the accelerating voltage V_a and H the strength of the magnetic field). By adjusting these two deflections to be equal, so that there was no net deflection of the spot on the oscilloscope screen, the charge to mass, e/m, for the electron was determined.*

known as the proton, involved in the electrolysis of water, while the magnitude of the charge in both cases was the same. He thus concluded that the cathode ray must involve a negatively charged particle (he used the term corpuscle) which was 1700 times lighter than hydrogen. Indeed he had discovered the first subatomic particle which is now known as the electron. Parenthetically it is an interesting fact that the original name for electrons, cathode rays, has survived to this day so that the video tubes in television sets and word processors are still familiarly known as CRTs. The fact that the electron is negatively charged means that the atom, which is normally electrically neutral, must contain an equal amount of compensating positive charge. In fact at the start of the twentieth century, the prevailing view of the atom, due to J.J. Thomson, was that it was comprised of some sphere of positive charge in which the negative electrons were embedded, the so-called plum-pudding model.[12] At this time another giant, Ernest Rutherford, a student of J.J. Thomson, came on the scene. He and others studied the byproducts of radioactivity discovered by Becquerel and named them alpha, α , beta, β, and gamma, γ , rays. Today the α and the β rays are both referred to as particles; the α-particle being a doubly positively charged helium atom that has been stripped of two electrons, whilst the β-particle is the electron. γ-rays are now known to be electromagnetic waves with even shorter wavelengths than x-rays.[12]

The weight of the proton having been determined to be 1.67×10^{-24} grams and knowing the density of copper to be 8.92 grams per cubic centimeter, it is easy to calculate that there are 8.9×10^{22} atoms in one cubic centimeter of copper. If the atoms are uniformly spaced spheres (assumed of negligible size for the moment) then the average distance between each atom is about 5 Å. Thus in a solid, which is the densest form of matter under normal temperature and pressure, the interatomic distances are comparable to the wavelength of x-rays. This fact was noted by von Laue who therefore suggested that a crystal would act as a diffraction grating for x-rays.[21] This was based on the supposition that in crystals the atoms are arranged like automobiles in a multi level garage in an ordered three-dimensional geometric arrangement or lattice. The existence of such geometric order had been inferred from the occurrence of the geometric growth and symmetry of natural or synthetic crystals. The predicted diffraction phenomena, first observed by this Austrian physicist in 1912, thus provided experimental verification of the lattice model of crystalline solids.[21] Shortly thereafter x-ray diffraction was used to study the structure of crystals and measure precisely absolute interatomic distances. For copper, the interatomic distance of about 2.5 Å is the size of a typical atom.

It was Rutherford's genius to use the positively charged α-particle as a bullet to shoot at other atoms. His experiments, reported in 1911, were simple in concept and involved firing a beam of α-particles at a thin metal

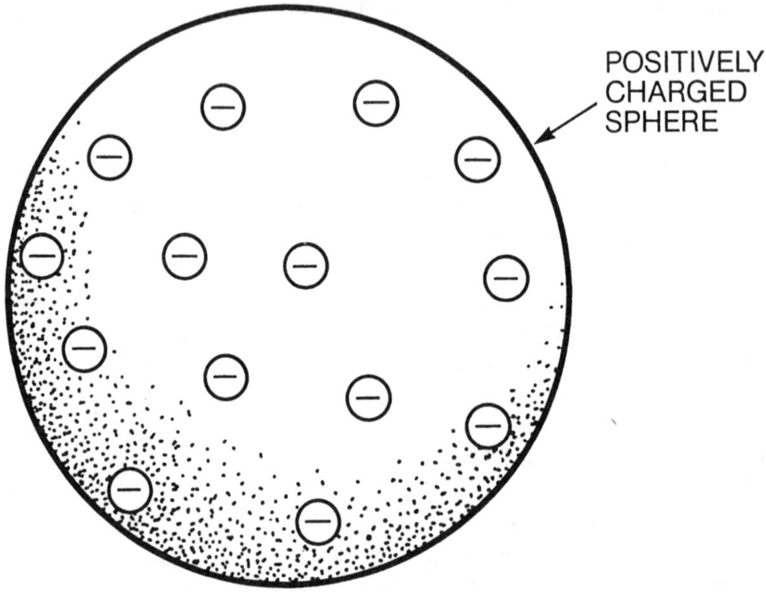

The so-called plum-pudding model of the atom proposed by Thomson after his discovery of the electron. Since atoms are neutral, the negative electrons have to be associated with an equal amount of positive charge and were visualized as being like raisins distributed in a spherical plum-pudding. This model preserved the indivisible character of the atom proposed by Democritus and championed by Dalton.

foil and detecting what happened to them using a scintillation detector. The latter consisted of a window coated with zinc sulfide which produces a flash of light when struck by an α-particle. Rutherford reasoned that the α-particles, even though charged, would go through the foil with no effect, since in Thomson's atomic model the atom would be an electrically neutral sphere. Contrary to expectations, Rutherford was amazed to find that some of the α-particles were being deflected as if they had encountered some enormous repulsive electrical force. Rutherford's conclusion was that the atom must have a structure in which a positively charged nucleus containing most of the atom's weight is surrounded by negatively charged electrons rotating in orbit, like planets around a sun. Despite the fact that effectively all the mass of the atom is contained in the nucleus, the diameter of an atom is roughly 10^5 times larger than that of the nucleus. Atoms truly are mostly empty space.[12,19]

Today this is a commonly used pictorial visualization of an atom. Unfortunately as Rutherford himself fully appreciated, there was a totally unacceptable consequence of this essentially classical picture. Any body in

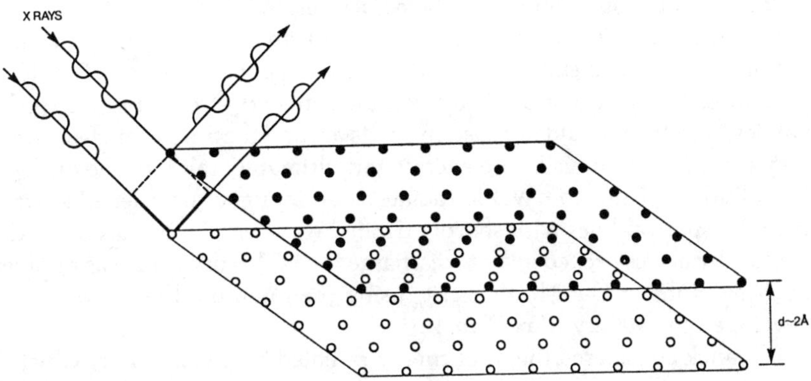

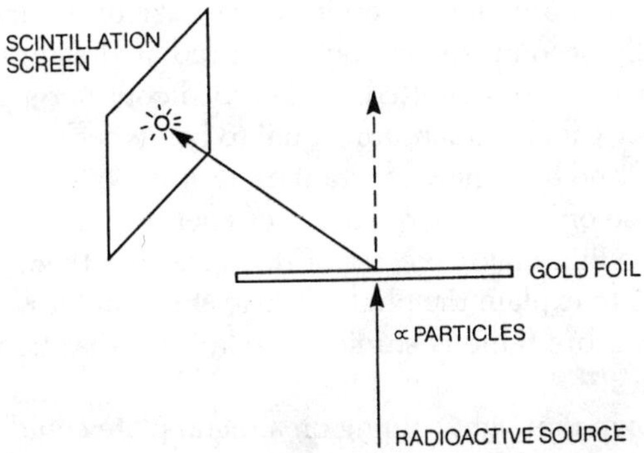

Top: *The periodic arrangement of atoms on a crystal lattice, initially deduced from the symmetry of naturally occurring crystals, is nature's way of providing a regular array of diffracting planes. Their separation is only a few angstroms so that only electromagnetic waves with wavelengths much shorter than visible light can produce diffraction. The details of the diffraction pattern depends on the different distances (thick lines) traveled by rays which strike, for example, the first and second plane of atoms respectively. With the discovery of x-rays, diffraction from crystals became possible. This confirmed both the short-wavelength character of x-rays and the periodic lattice structure. From an analysis of the diffraction patterns, measurements of inter-atomic distances are possible.* Bottom: *The experiment by Rutherford that finally revealed the structure of atoms. Positively charged alpha, α, particles, derived from radioactive materials, were fired at a thin film target. On the basis of the plum-pudding model of an overall neutral atom, they were expected to be largely unaffected. Instead, unexpectedly large deflections were detected as flashes of light as they struck a scintillation screen. Rutherford was thus forced to suggest the existence of a positively charged nucleus, containing most of the mass of the atom, surrounded by very light electrons of sufficient number to provide the required compensating equal negative charge.*

a closed orbit around another is being accelerated. It was a basic tenet of the electromagnetic theory of Clerk Maxwell that an accelerating charge radiates electromagnetic waves, and thus inevitably loses energy.[12,13] Thus according to the classical theory of motion and electricity (electrodynamics), electrons could not sustain orbits around the atomic nucleus, since they would continuously lose energy and ultimately fall into the nucleus. In addition such an atom would radiate a continuum of wavelengths rather than the spectral lines with sharply defined wavelengths characteristically emitted by atoms excited in a gas discharge tube. This dilemma was resolved by Niels Bohr, a Danish physicist, using the quantum theory ideas first developed in 1900 by Max Planck.[22]

Planck considered the total energy radiated by a black-body which is a cavity in a block of material connected to the outside surface by a small hole and approximates to a perfect emitter and absorber of radiation, hence the name black-body. The radiation emerging from such a cavity depends only on its temperature and not on its shape, size or nature. Planck could only analytically account for the observed radiated energy if he assumed that radiant energy is not emitted or absorbed continuously but only in a discrete quantity called a quantum, equal to $h\nu$ where ν is the frequency of the radiation and h is a new constant of nature, now called Planck's constant. In the case of light, the quantum of energy is given a special name, the photon. The first major success of this quantum theory was its use by Einstein in 1905 to explain the photoelectric effect and it was for this work rather than his more famous studies of relativity that he won the Nobel prize for physics.[23]

It was known that light shining on a metal plate could cause electrons to be emitted from it. Unexpectedly and contrary to classical wave theory, the maximum energy of these electrons was found to be independent of the light intensity and determined instead by its frequency or photon energy. A situation was thus emerging in which light behaved as a wave in that it exhibited interference and diffraction phenomena, whilst other effects such as black body radiation and the photoelectric effect seemed to imply that it was a particle. The modern view is that light possesses both wave and particle-like properties and can play a variety of roles. However this does not mean that light should be thought of as a classical wave in which the square of the wave amplitude determines its energy. The amplitude of a light wave is related to the probability of finding a photon of a given energy at a given point in space.[24,27] In the photoelectric effect, for example, the number of photoemitted electrons and thus the photocurrent is proportional to the intensity of the incident light while in an interference experiment bright lines correspond to an enhanced probablity of finding a photon at that position and dark lines as occurring due to a reduced probability. For the purposes of propagation, light can, as Newton insisted, be thought

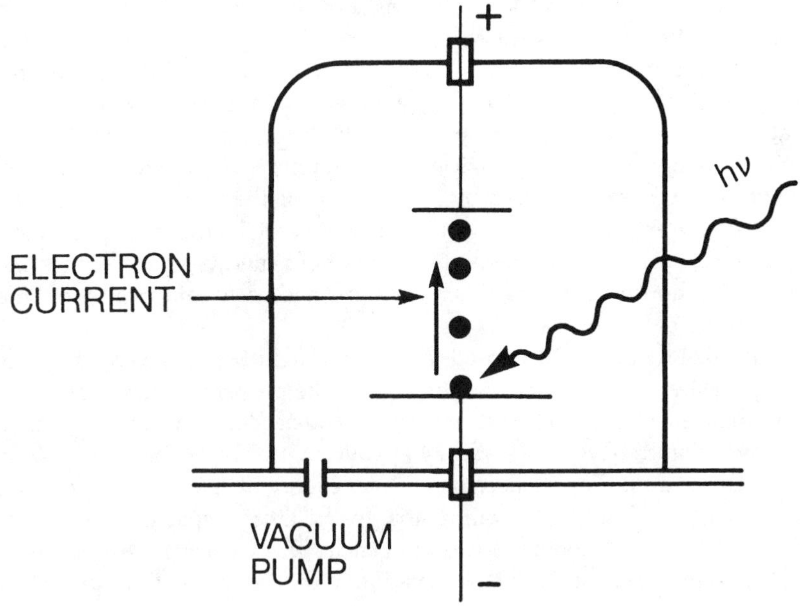

The photoelectric effect involves the ejection of electrons from an illuminated surface such as that of a metal. In a vacuum, these electrons can be drawn to a positively charged anode constituting a current flow. By applying a negative voltage to the anode sufficient to suppress the electron emission, the electron energy can be measured. Under classical physics, this would be expected to be proportional to the intensity of the light; in reality, it is determined by the frequency of the light. It is the magnitude of the photocurrent which is determined by the light intensity and vacuum-tube photodetectors or photomultipliers used to measure light intensities work on this principle.

of as a small projectile carrying a packet of energy determined by its frequency and thus the need of a medium for propagation does not exist. This is reassuring, for despite extensive and ingenious experimental efforts, the hypothetical ether has never been discovered.

Bohr resolved the dilemma of the atomic structure by postulating that electrons in an atom can occupy orbits within which they can have discrete and constant energy.[24] These were called stationary states. Energy can only be gained or lost by an electron moving between stationary states of different energy. Under such conditions, a photon of energy $h\nu$, equal to the energy difference between the states, will be emitted or absorbed. For an essentially isolated atom, approximated by a low pressure gas discharge, the optical emission spectra would consist of sharp lines reflecting electron transitions between discrete energy states. While Bohr's ideas passed the ultimate test of explaining the otherwise inexplicable sharp atomic spectra, they were incomplete in that they gave no explanation how stationary states could exist but only the postulate that they did. A commonly used unit of energy for atomic particles and photons is the electron volt, eV, which is the energy acquired by an electron, or the work which must be done on an electron, in moving it through a potential difference of one volt. An electron volt is an extremely small amount of energy, since it equals one million millionth of an erg where one erg has been compared with the softlanding of a snowflake.

In 1924 de Broglie proposed that, just as light behaves as both waves and particles, all matter might have wave-like properties.[25] Accordingly, any particle with mass, m, and velocity, v, and hence momentum, p, where $p=mv$, would have an associated wavelength, $\lambda = h/p$, where h is Planck's constant. For an electron with an energy of 100 eV, the predicted wavelength is about 1 Å, comparable to the atomic spacing in crystals. Consequently the wavelike nature of particles was confirmed, by Davisson and Germer in 1922, with the observation of diffraction patterns produced by firing a beam of electrons at a crystal. An obvious question concerns the nature of the waves associated with an electron, which under some circumstances clearly behaves as a particle. This was resolved and Bohr's revolutionary picture of the atom refined by the development of quantum mechanics and more specifically by wave mechanics introduced by Schrödinger in 1926. The central feature of this new theoretical description is that an electron, like light, has both particle and wavelike properties; the so-called wave-particle duality.[26] The waves, as in the case of light, are probability waves and the square of the wave amplitude describes the relative probability of finding the particle in different regions of space. The reconciliation of this with Bohr's picture of stationary orbits lies in the properties of wave motion. Earlier it was pointed out that musical string instruments function because the frequencies of waves possible in the string of fixed length are constrained. It turns out that Bohr's stationary orbits are precisely those in which standing waves can exist. For this to be possible, the circumferences of such orbits have to be integer multiples of the electron wavelength. But if the particle wavelength is restricted, so is its energy. Hence, as postulated by Bohr, there are only certain allowed orbits and therefore energies for electrons in an atom. However, the electrons can no longer be regarded as physically moving about the central nucleus of the atom. As a result of the Heisenberg Uncertainty Principle, the energy and position of the electron can never be exactly known simultaneously. A more accurate, but intrinsically incomplete picture, is of a nucleus surrounded by electron clouds representing the probability of finding the

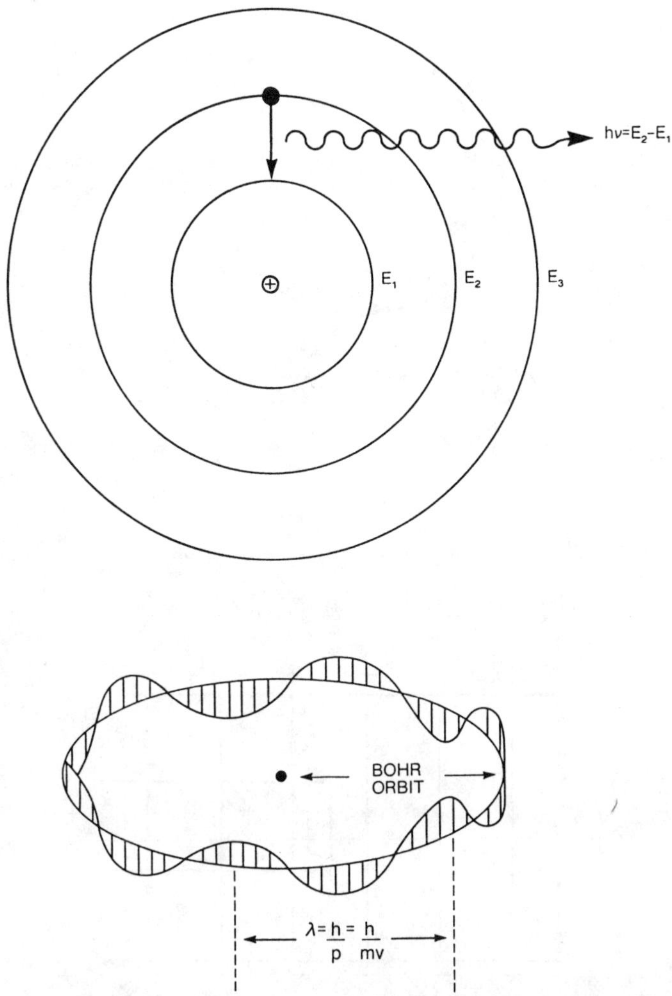

The planetary picture of the atom due to Bohr who postulated that the electrons occupied stationary or constant energy orbits surrounding the positively charged nucleus. An electron in orbit with the electron energy, E_2 *can only change its energy by moving to higher energy orbit,* E_3*, or a lower one,* E_1*. If an electron is given sufficient energy it can escape completely from the atom which is said to be ionized. In the transition from* E_2 *to* E_1 *a quantum of radiation* $h\nu$ *(h is Planck's constant and* ν *is the photon frequency) equal to the energy difference* E_2*-*E_1 *is emitted. In isolated atoms, such as in a gas, the orbital energies are sharply defined, accounting for the line spectra observed in gas discharges. A theoretical justification for the stationary orbit postulate depends on the wave-particle duality of matter. An electron with a mass m and velocity v and therefore momentum,* $p=mv$ *has an associated wavelength* $\lambda=h/p$*. The condition for a stationary orbit is that an integer number of such electron wavelengths must just fit within the circumference of such an orbit.*

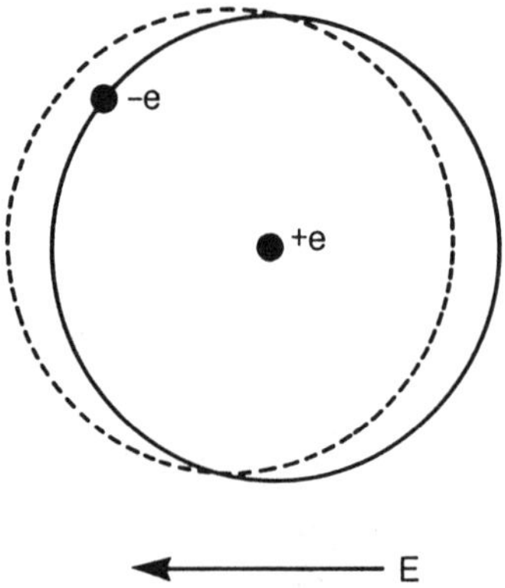

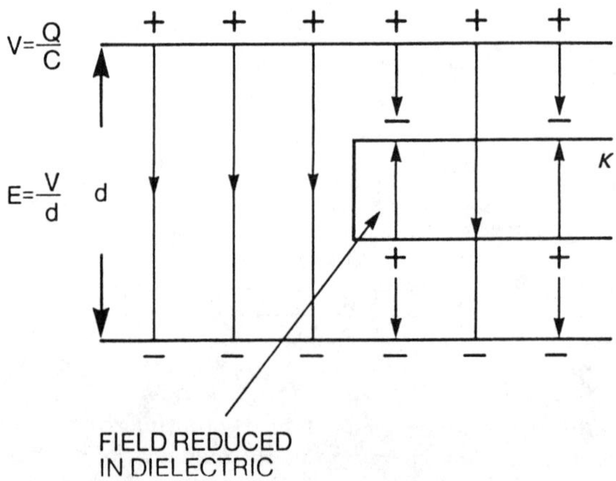

The dielectric constant of matter is a consequence of the atomic structure. Normally the electron clouds are symmetrically distributed about the positively charged nucleus. With an applied electric field, the electron distribution is shifted asymmetrically so that a net surface charge of negative polarity appears adjacent to the positive voltage and a positive surface charge adjacent to the negative voltage. Thus the capacity of a capacitor is enhanced by the amount of this induced surface charge which is material specific and determines its dielectric constant, κ*. The field induced in the dielectric slab is opposite to that due to the voltage on the capacitor electrodes and causes a reduction in the field, inversely proportional to* κ*, within the solid.*

electron. These probability clouds are densest at the radii of the orbits predicted by Bohr.[27]

The dielectric constant of materials and its effect on the storage capacity of capacitors can be understood in terms of the atomic structure because of what is known as the polarizability of matter.[12] When a voltage is applied to a material, the center of the electron cloud distribution in each atom is shifted slightly with respect to the positively charged nucleus. As a result, although the material as a whole remains electrically neutral, the net effect is for excess negative charge to appear at the surface adjacent to the positive voltage and excess positive charge on the opposite surface. This excess charge which determines the dielectric constant is the extra amount that can be stored. Since the induced polarization charges are opposite in sign to the applied voltage, they act to reduce the applied external electric field.

The total number of electrons in an atom equals the positive charge on the nucleus which in turn equals the atomic number and thus identifies the element. However the various electron orbits in an atom can only be occupied by a limited quota of electrons. Thus moving outward from the orbit nearest the nucleus, the different orbits can progressively accommodate 2, 8, etc., electrons. A hydrogen atom, for example, consists of a single electron occupying stationary states around a singly positively charged nucleus. The energy to remove this electron from the lowest energy or ground state of a hydrogen atom to produce the positively charged hydrogen ion, called the proton, is 13.6 eV. Generally speaking, it is the last occupied or first partially occupied levels that are of significance in terms of bonding in solids, chemical reactivity and electrical phenomena. If the last occupied orbit is filled with its allotted quota, then the atom is said to have a "closed-shell" and is exceptionally inert, so that elements with closed shells such as helium, neon, argon or krypton are chemically inactive and called the inert gases. By contrast elements whose last occupied orbit does not have its full quota such as sodium or chlorine are highly chemically reactive.[24,28]

So far the discussion has been limited to the structure of individual atoms. In order to explain the electrical and photoelectrical properties of solids, attention must now turn to how atoms, which are the building blocks of matter, interact with each other to form chemical compounds and specifically solids. Unlike gases and liquids, solids maintain their own shape because they are held together by specific bonding forces between their constituent atoms.[34] Three types of bonding, ionic, covalent and Van der Waals, can exist in solids and give different classes of solids some of their characteristic properties; in addition to the type of bonding other aspects such as the bond lengths and angular or stereo-projections into three-dimensional space are also important. The best example of the ionic

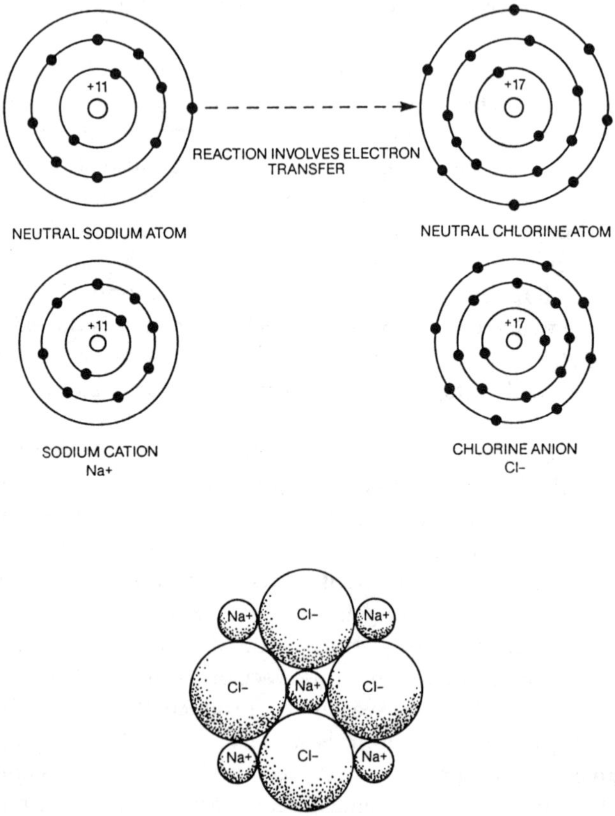

The bonding together of atoms largely determine the physical properties of solids. The ionic bond is formed when electrons are transferred from one atom to another to achieve closed-shell configurations. The cohesive force in such an ionic solid is the strong coulombic force between the oppositely charged ionic building blocks. Electrical conduction in such solids at room temperature is due to the motion of ions as they drift into vacant lattice sites present even in the most perfect crystals man can grow.

bond is that formed between an alkali metal such as sodium and a halogen such as chlorine. Sodium has an outer shell containing a single valence electron, whereas chlorine has seven. Thus when sodium and chlorine chemically react they can each achieve a stable closed-shell configuration if each sodium atom gives its lone electron to a chlorine atom. A bond involving direct exchange of one or more electrons between the constituent

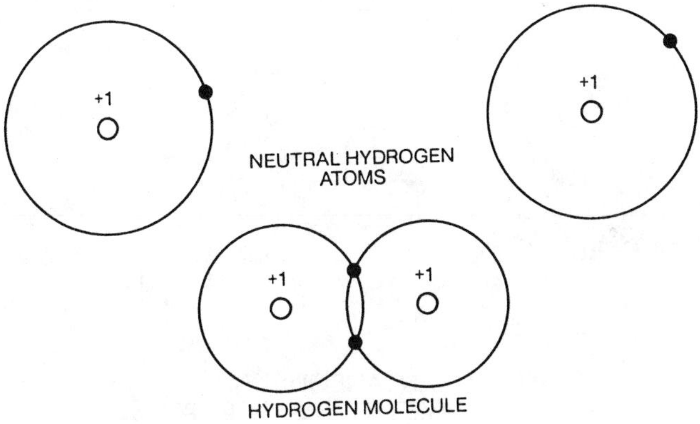

For other atoms, an example being hydrogen, the closed-shell is achieved by the sharing of one or more electrons; this is the so-called covalent bond. In addition to the hydrogen molecule, silicon and selenium are examples of covalently-bonded solids.

atoms of a solid is called an ionic bond. As a result of such an exchange, the sodium becomes a positive ion or cation and the chlorine a negative ion or anion. The cohesion in sodium chloride crystals, common table salt, therefore, is due to the coulomb attractive force between these oppositely charged ions. The fact that ions are the building blocks of salts explains why salt solutions are good conductors of electricity and are used as electrolytes in batteries.[17]

The hydrogen atom with its sole valence electron requires an additional one to achieve a closed-shell configuration and the hydrogen molecule, consisting of two bonded hydrogen atoms, is formed by the atoms sharing their two valence electrons. Such bonding involving the sharing of electrons is called covalent bonding.[17] Silicon is an element with four valence electrons in the outermost occupied shell of its atom. In an ideal silicon crystal, each of these electrons is involved in a chemical bond so that each silicon atom achieves a stable closed shell configuration consisting of eight electrons by sharing the four valence electrons of four other silicon atoms.[30,34]

If an atom already has a closed-shell, for example an inert gas such as neon, then neither a covalent nor ionic bond can be formed. Instead, the bonding in solid neon results from a more subtle quantum-mechanical force. Pictorially it can be understood by recognizing that, although on the average the negative electron cloud in an atom is distributed with spherical

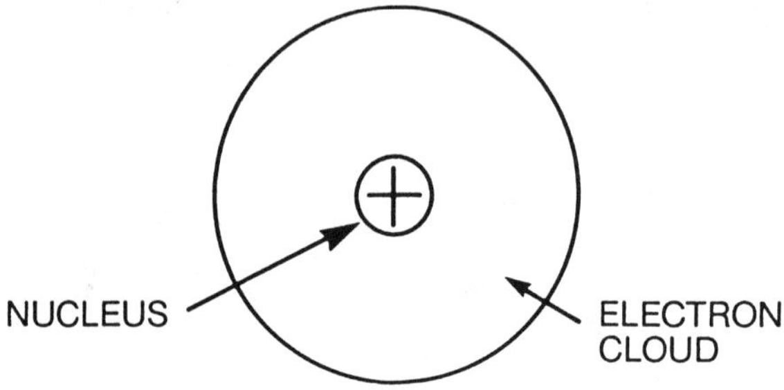

INSTANTANEOUSLY
SYMMETRIC NEUTRAL ATOM

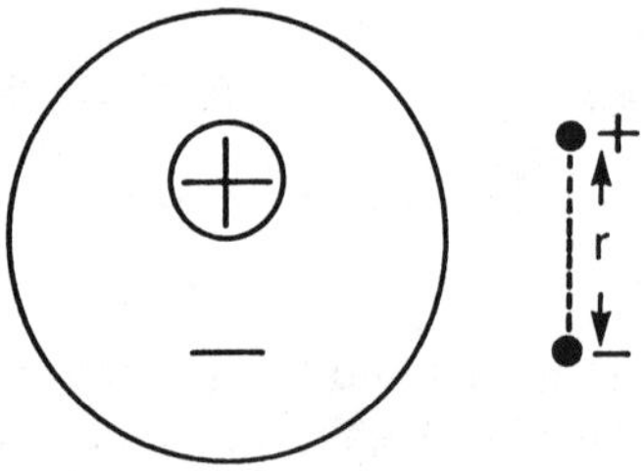

INSTANTANEOUSLY
ASYMMETRIC ATOM
YIELDING DIPOLE MOMENT

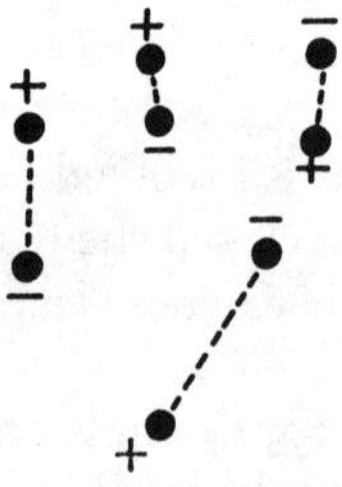

DISPERSION FORCE DUE
TO INTERACTIONS OF FLUCTUATING
DIPOLES

symmetry about the nucleus, this is not true at every instant. As a result, the center of the electron cloud may not instantaneously coincide with the nuclear center of the atom. In such circumstances, the atom has a spatially asymmetric distribution of charge and is said to have a dipole. It is the interaction between the fluctuating electromagnetic fields caused by such atomic dipoles in all the neon atoms that are the source of the cohesive energy in neon crystals. The dipole interactions can be attractive or repulsive depending on the relative alignment of the dipoles. However, on the average the interaction will be more attractive since this represents the lowest energy state. Generally this is a relatively weak force as manifested by the fact that solid neon melts at minus 248°C.

As a consequence of these advances, the relationships between the structure of atoms and electrical phenomena were reasonably well understood for the case of liquid solutions and gases[17,28,29] where, based on this insight, the conducting species are ions produced by dissolving salts or in the case of conduction in gases involving the creation of ions by irradiating the gas with x-rays or by applying very high electric fields.[29] The latter is the method used to charge photoreceptors in xerography, while the phenomena of triboelectricity involves the addition or removal of excess charge induced by friction between two solids.

The understanding was less satisfactory in the case of the electrical conductivity of solids, where no fundamental explanation of the distinctions between metals and insulators existed. There was of course a clear recognition from the earliest days of man's history that there was a class of solids which were readily characterizable, namely metals, as typified by the "Bronze and Iron Ages." Even before the discovery of their ability to conduct electricity, metals such as gold and silver were appreciated for aesthetic visual qualities deriving in part from their extremely high reflectivity of light; a characteristic property of metals. More plentiful and practical metals such as iron and copper had an exceptional ability to conduct heat, as contrasted to stone, and their malleability, ductility and strength had for centuries led to their use as utensils and weapons.[10] It is a

Opposite: *Atoms in solids like inert gases already have closed-shells so that neither ionic or covalent bonding can exist. While on the average an atom is neutral, since the electron distribution is symmetrically distributed around the nucleus, instantaneously this distribution can be asymmetric. This results in the formation of an electric dipole as a relative displacement of the positive and negative charge by a distance, r, occurs. For such solids, the mutual interaction of such instantaneous, fluctuating dipoles provides the cohesive force. This Van der Waals or dispersion force is much weaker than the ionic or covalent bonds. Organic solids are typically composed of molecules bonded by such forces, although within the molecules covalent or ionic forces bond the atoms. This type of quantum-mechanical force is vitally important in the adhesion between solids and plays a critical role in xerography.*

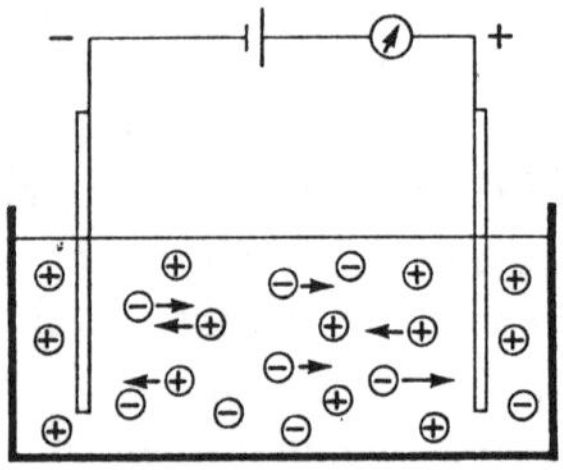

SALT SOLUTION

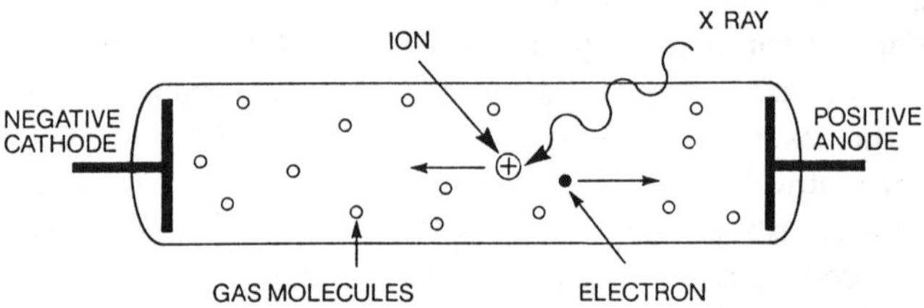

GAS DISCHARGE

Ions are atoms with either a deficit of electrons, (positively charged and known as cations since they move to the cathode), or atoms with an excess number of electrons, (negatively charged and called anions). Conducting ionic solutions are formed by the dissolution of solids which are themselves formed of ions; common salt being an example. In gases, the ions are formed by bombardment with energetic α or β particles or x-rays.

remarkable fact that it was not until the mid-nineteenth century that serious systematic studies of the electrical properties of solids were begun. This, of necessity, was preceded by the developments in the understanding of electricity and the means of its production. For further progress to be made, particularly in understanding the electrical properties of solids on a microscopic level, there had to be parallel efforts in materials processing.

At the beginning of the nineteenth century, the situation was that there were excellent conductors of electricity known, these being the metals, and exceedingly poor conductors, the insulators. Scientists use a number of terms to define the facility with which a material can conduct electricity.

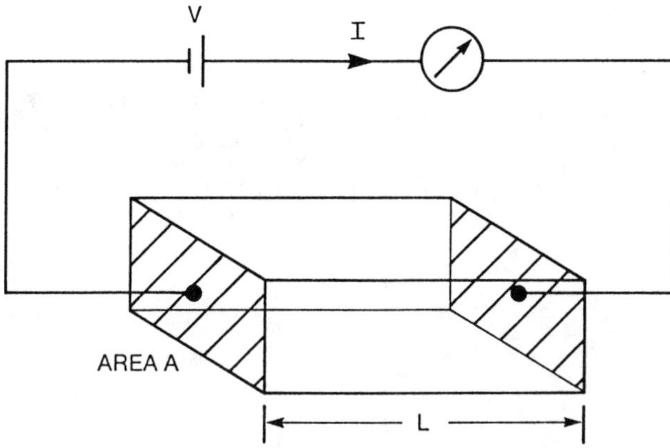

The resistance to electrical conduction of a material, R, given by Ohm's law, as the applied voltage, V, divided by the current flow, I, it produces, is determined by the geometrical dimensions of length, L, and area, A, of a particular sample. The resistivity, ρ, given by RA/L, eliminates such geometrical factors and becomes material specific (the conductivity, σ, is equal to 1/ρ).

Perhaps the most familiar is resistance which, as the word implies, is a measure of how a material inhibits the flow of current when a voltage is applied to it.[12] Thus if 1 volt is applied to a solid and a current of 1 ampere flows then the solid is said to have a resistance of 1 ohm or 1 Ω. This is the simple application of Ohm's law which states that the applied voltage, V, equals the product of the current, I, flowing through a resistance, R. Thus a 100 watt light bulb operating on a 110 volt supply has a resistance of about 120 Ω with a current flow of 0.9 ampere. More typically a parameter termed resistivity, with units of (Ω cm), is used because this is a measure of the same property, except that it is independent of the geometric dimensions of the solid and therefore more fundamental.[12] Conductivity is the inverse of resistivity so that metals have a high conductivity or equivalently a low resistivity. For the purposes of discussing xerography, it is more convenient to talk in terms of resistivity. The range of electrical resistivities in solids is extraordinarily large being about 10^{-6} (Ω cm) for metals to 10^{14} (Ω cm) or larger for the best insulators.[28]

Investigations started in the nineteenth century revealed that certain substances, originally called "poor" conductors, could not be easily

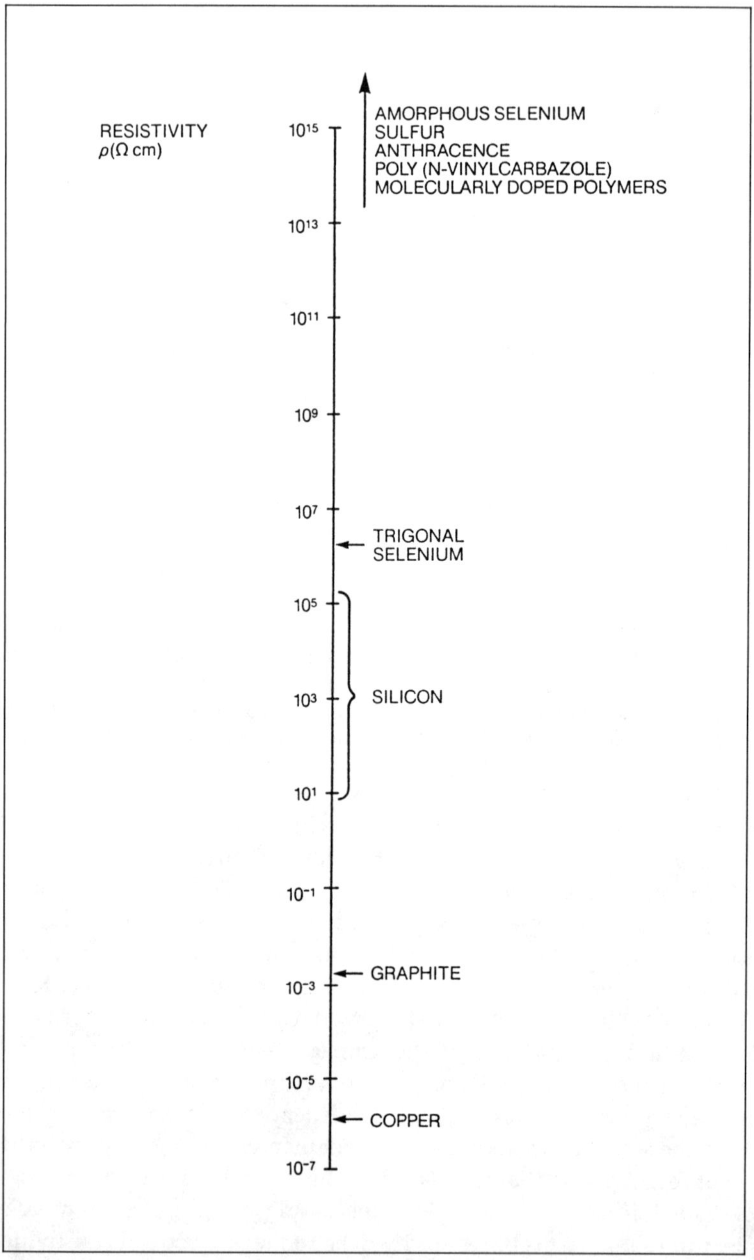

The range of resistivities observed in solids is enormous and ascending the scale shown the resistivity increases by a factor of 100 for each division. Values for representative classes of materials including those important in xerography are indicated.

categorized as either metals or insulators. A major problem for the early experimenters, and it was almost totally an experimental activity, was the question of the purity of the materials being studied. It is only with the perspective of the modern day understanding of the solid-state that the difficulties of these early pioneers can be appreciated. Despite this, it is a testament to their skill that so much progress was made in recognizing the "poor" conductors as a new class of solids, known today as semiconductors, long before a rigorous theory was available to explain and account for them.[30] Indeed, the experimental observations amassed during the nineteenth and early twentieth centuries played a critical role in the development and ultimate test of these theories. Semiconductors usually have intermediate resistivities in the range 10^{-3} to 10^{6} (Ω cm).[28,30] As a general, but not absolute, rule semiconductors exhibit a negative coefficient of resistance, i.e., their resistance to the flow of electricity falls as the temperature rises; for metals it increases.[30] Michael Faraday in 1833 first noted the extraordinary case, as he called it, of silver sulfide which exhibited a conducting power comparable to a metal only when heated. Somewhat prophetically, he suggested that, if sought, similar substances would be found.[14] Subsequent to Faraday's observation, little new progress was made, except for the verification of his prediction of finding more semiconductors, until 1873 when the phenomenon of photoconductivity was discovered for the first time.

Photoconductivity, the enhancement of the electrical conductivity of a material on its exposure to light, played a special role in twentieth-century development of the physics of solids and is integral to xerography. Historically it was first reported by Willoughby Smith, an electrician for a telegraph company. By one of those curious coincidences, his observation was made in the element selenium which would prove so important in the development of xerography.[31,32] Smith, using rods of selenium as resistors in the testing of submarine cables, found that their resistance changed depending on whether they were enclosed in a light-tight box or exposed to a light source. He eliminated the other possible cause for a resistance change, namely, a change in temperature of the rods caused by the light sources used, such as a gas burner or burning magnesium; very few light bulbs existed in those days. Interestingly enough Smith described the selenium he used as a high resistivity metal, which exemplifies the uncertain situation at the time with respect to the classification of solids.

In 1879 it was observed that a voltage could be developed in a current-carrying conductor situated in a magnetic field. This effect was called the Hall Effect after the scientist who first observed it.[33] Eventually this was to prove a critical diagnostic tool for distinguishing between electronic and ionic conduction and between metals and semiconductors. At the time of its discovery there was no firm basis for either understanding or appreciating its significance, since it even predated the discovery of the electron.

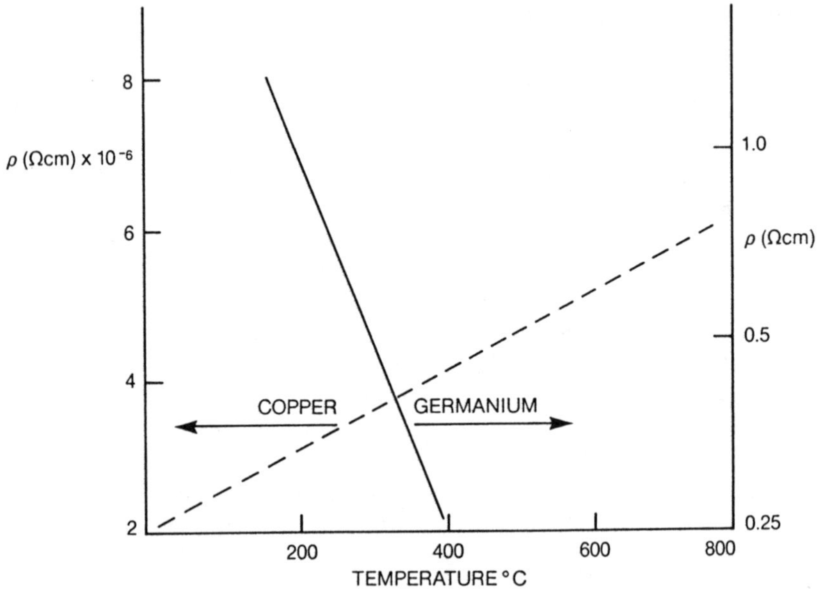

The electrical resistivity of a typical metal such as copper changes much less with temperature and in the opposite sense to that of the semiconductor, germanium. Even at the highest temperature, however, germanium is still about a million times more resistive than copper.

Three fundamentally characteristic properties of crystalline solids, namely atomic bonding, periodicity and their highly condensed nature coupled with quantum mechanics, finally resolved the major mysteries concerning their electrical properties.[28,30,34] The atoms in a crystal, bonded to each other by either ionic, covalent or Van der Waal forces, are arranged on a three-dimensional periodic lattice that extends effectively to infinity; 1 cm$=10^8$ Å is a good approximation to infinity relative to the atomic scale. Moreover the constituent atoms are packed so closely together that the electronic clouds associated with adjacent atoms strongly interact with each other. For example, two atoms which are initially very far apart have the same allowed energy levels. As these atoms are brought close to each other, the electrons interact by mutual coulomb repulsion and an identical energy level changes into two; one higher and one lower than the one in the isolated atoms. With N atoms in a crystal, the initial level splits into N separate levels. This is true for those electrons furthest from the nucleus, the so-called valence electrons. Deeper lying levels, appropriately called

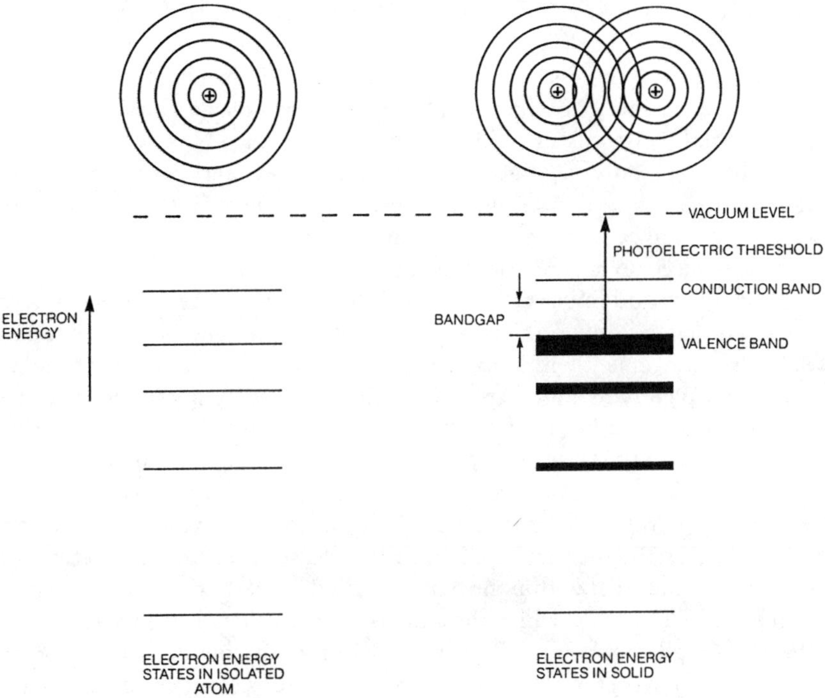

The basic characteristic feature of solids is that about 10^{22} atoms are contained within one cubic centimeter so that the interatomic spacing is only a few angstroms. Adjacent atoms, therefore, interact through a mutual repulsive coulomb force between their electron clouds. The degree of interaction and resultant broadening is greatest for electrons in the outermost orbits, since the innermost orbits are screened and remain atomiclike. The discrete energy levels of an isolated atom are thus broadened into bands of allowed and forbidden energies for the electrons; the periodic nature of the crystal lattice results in these energy bands being a property of the solid. The dashed line indicates the vacuum level which represents the energy at which the electron becomes free of the atom or solid. The photon energy indicated by the arrow, which gives an electron in the valence band this energy, corresponds to the threshold frequencies observed in the photoelectric effect. For the photoelectric effect of a metal, a related energy called the work function is used. This is the basis of Carlson's first invention and U.S. Patent in electrophotography which he called Electron Photography.

core levels, are screened from, and much less perturbed by, neighboring atoms and retain the discreteness typical of an isolated atom. In this simplified description, the discrete outer electron energy levels of an atom in a solid are broadened into bands of allowed energies several eV wide. Each band actually consists of N discrete levels but, since N is of the order of 10^{22}, these are only separated by about 10^{-22} eV and so the bands are essentially a continuum of allowed energies. However just as in the atom where there are nonallowed electron energies, in a crystal there is a range of forbidden electron energies between the allowed bands, called the bandgap, E_g. The upper and lower extremes of each band and their occupancy with electrons depend only on the elements of which the solid is comprised and on their arrangement within the crystal. This modern band theory of solids provides a picture that answers most of the outstanding questions regarding the electrical and photoelectrical properties of solids.[28,30,34]

Initially only the uppermost energy band which is occupied need be considered. Two situations are possible, either this band is partially filled or completely full as a consequence of the nature of the valency of the constituent atoms and the bonding. In the former case, a continuous distribution of vacant states exist within the partially filled band, known as the conduction band. The level to which the band is filled is called the Fermi level after the Italian physicist Enrico Fermi who is perhaps better known for his assembly of the first nuclear reactor.[28,30,35] Electrons in the conduction band cannot be considered to belong to any particular atom, rather they are viewed as being delocalized throughout the crystal. These delocalized electrons can be promoted to unoccupied levels within the conduction band with small increments of energy. Such a solid is what we call a metal. When an electric field is applied to a metal these free, unbonded electrons are accelerated in the direction of the electric field which supplies them with the necessary energy to move to higher energy states within the partially-filled band. This constitutes the electric current flow which is the rate of flow of charge and equals the product of the total electronic charge moving times its speed. The speed of the electron is proportional to the applied electric field; the higher the field, the faster the electrons move and the larger the current. The speed per unit field, called the electron mobility, is a useful measure of the facility with which an electron moves in a particular material.[28,30] Since each electron carries the electronic charge, e, the parameters which determine the electrical conductivity or resistivity of a material are n, the density of free electrons per unit volume and their mobility, usually denoted by μ; in fact the conductivity $\sigma = ne\mu$. In a normal metal n is constant, not changing with temperature, and is $\sim 10^{22}$ cm^{-3} and the mobility $\mu \sim 30$ cm^2/Vs and decreases with increasing temperature. This accounts for the characteristic decrease in conductivity of a metal with higher temperatures.

At the other extreme is the case where the last occupied band is completely full. In this case the electrons cannot move to any higher allowed energy states within this band, known as the valence band, because there are none vacant. In principle it is possible for an electron to move from the valence band to a vacant state in the next band which is an empty conduction band. However to do this it would have to gain an energy equal to or greater than the bandgap. One possible source of this energy, thermal, can be supplied by increasing the temperature of the solid. For some solids the energy gaps are greater than 5 eV while, by contrast, the thermal energy associated with an electron at room temperature is only 0.025 eV. The probability of thermal excitation is proportional to the negative exponential of the ratio of the bandgap to this thermal energy.[28] As a result the thermal excitation rate of electrons at room temperature is completely negligible for such wide bandgap materials which are known as insulators. Thus the band theory of solids gives an explanation of the basic difference between a metal and an insulator, first recognized by Gray in 1729. In metals, the conduction band is partially filled with electrons whereas in an insulator it is empty even at room temperature. In fact, a metal can be viewed as consisting of a three-dimensional array of positive ions immersed in a sea of free electrons, whereas in an insulator no such sea exists. Thus the electron sea is the medium that Faraday was searching for in his paradox; the critical insight he lacked was the fact that the atom itself is divisible.[14]

Semiconductors are also explained by the band theory of solids. Conceptually they can be thought of as a special case of an insulator in that at very low temperatures the highest occupied band is totally filled. However the width of the bandgap is relatively small so that as the temperature of the material is raised, some of the electrons in the fully occupied valence band can thermally acquire the excess energy to move to the normally empty conduction band. The density of such electrons increases exponentially with increase in temperature and is also determined by the magnitude of the bandgap. Once in the conduction band, such freed electrons can move in an applied external field and an electric current flows. The higher the temperature, the higher the electrical conductivity becomes and can be significant at room temperature. This is the explanation for Faraday's observation in silver sulfide since, in contrast to a metal, the free electron density in a semiconductor is not constant but increases rapidly with temperature, although the absolute density is always much less than in a metal, while the electron mobility is still weakly temperature dependent. Semiconductors in which the electrical conductivity is determined by the density of carriers thermally excited across the bandgap are called intrinsic semiconductors.[28,30]

Since silicon is *primus inter pares,* electrical conductivity in semiconductors will be discussed from this vantage point. Intuitively one would expect the four neighboring silicon atoms to be arranged symmetrically with each silicon atom in the crystal. Nature achieves this by the silicon atoms

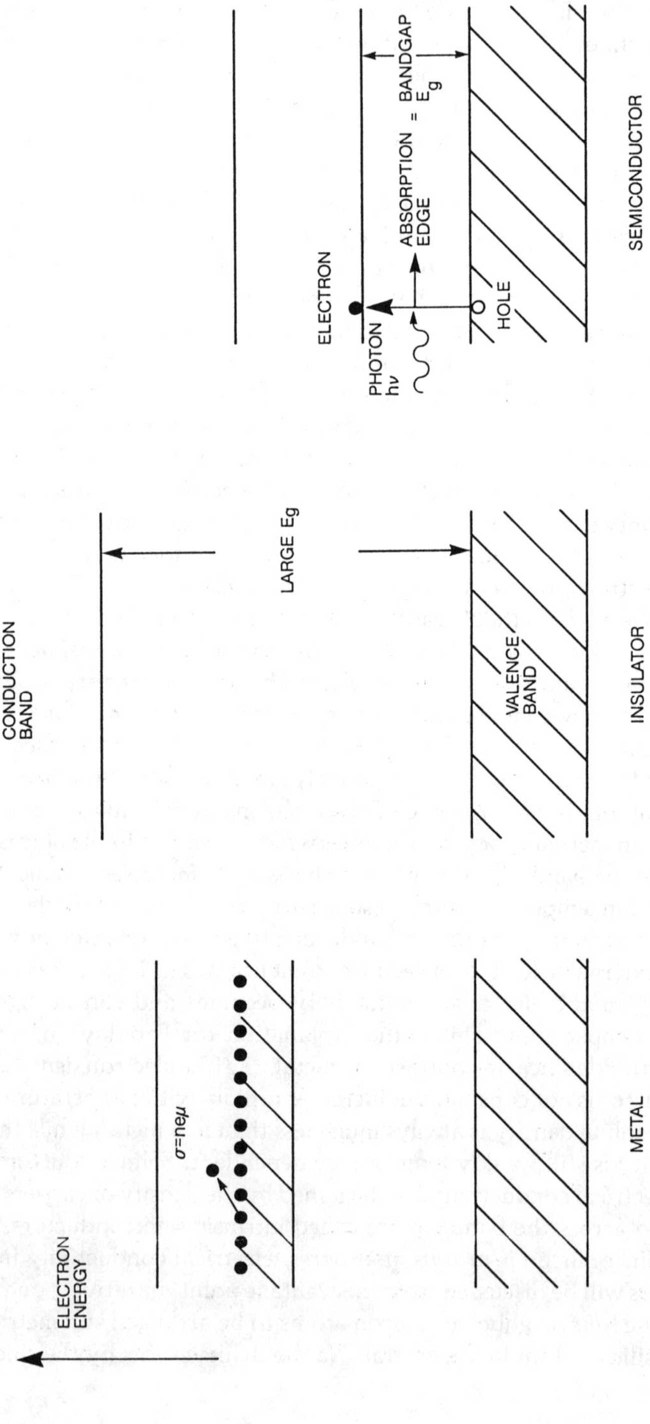
ELECTRON ENERGY
σ=neμ
METAL
CONDUCTION BAND
LARGE Eg
VALENCE BAND
INSULATOR
PHOTON hν
ELECTRON
HOLE
ABSORPTION EDGE
BANDGAP = Eg
SEMICONDUCTOR

occupying each corner and the center of a regular tetrahedron; the crystal lattice consists of a three dimensional array of this subunit. The four covalent bonds in silicon crystals account for all the valence electrons and form the fully occupied valence band. Thus at temperatures approaching absolute zero, such an ideal silicon crystal would be an insulator. However the bandgap of silicon turns out to be quite small, ~ 1 eV, so that at room temperature enough thermal generation across the bandgap causes silicon to behave as an intrinsic material. Physically this means that at room temperature there is sufficient thermal energy to allow a certain number of the valence bonds to be broken. The energy of the bandgap is that energy which must be provided to take such an electron from a bonding state and place it on a distant silicon atom. This forms a Si^- ion and the transfer of this excess electron from one nonbonding state of a silicon atom to another constitutes electron transport in the conduction band. The removal of a bonding electron gives rise to a Si^+ ion and the transfer of an electron from a neighboring neutral silicon atom corresponds to an effective positive charge moving through the bonding states in the opposite direction. This motion which occurs through the valence band is known as hole transport.[35,35]

The discussion of the band theory so far has been idealized in that the materials have been assumed to be crystals of absolute purity and perfection. Thus the material only consists of the atoms characteristic of the particular material and these occupy only the positions on the appropriate three-dimensional crystalline lattice. In the real world most materials

Opposite: *To understand the basic distinctions between the different categories of solid one need only consider the uppermost partially occupied or unoccupied band. In a metal the uppermost occupied band, the conduction band is only partially filled with electrons. These electrons, of constant density of 10^{22} per cc, each carrying a charge, e, can move freely with velocity, $v = \mu E$, (μ is the electron mobility and E is the applied electric field) in a quasi-continuum of unoccupied levels. The electric current density J, the total flow of charge per square centimeter, per second, equals $ne\mu E$. The electrical conductivities of metals are weakly temperature dependent and the highest of all solids. In an insulator, the uppermost band is empty and separated from the lower, fully occupied band, known as the valence band, by a energy or bandgap of several electron volts. In the absence of illumination, the probability of an electron acquiring sufficient thermal energy to go from the valence to the conduction band is almost zero. Without significant free carriers in the conduction band, no electrical conduction is possible and the solid is an insulator. Semiconductors are less exactly definable, since they are characterized by having a sufficiently small bandgap that the probability of thermal excitation of an electron from the valence to the conduction band is significant at room temperature. In their case, the number of free carriers in the conduction or valence band and consequently their conductivities, intermediate between those of insulators and metals, increase strongly with higher temperature. Photons with energies equal to or greater than the bandgap energy define the absorption edge and are strongly absorbed as an electron is excited from the valence to the conduction band.*

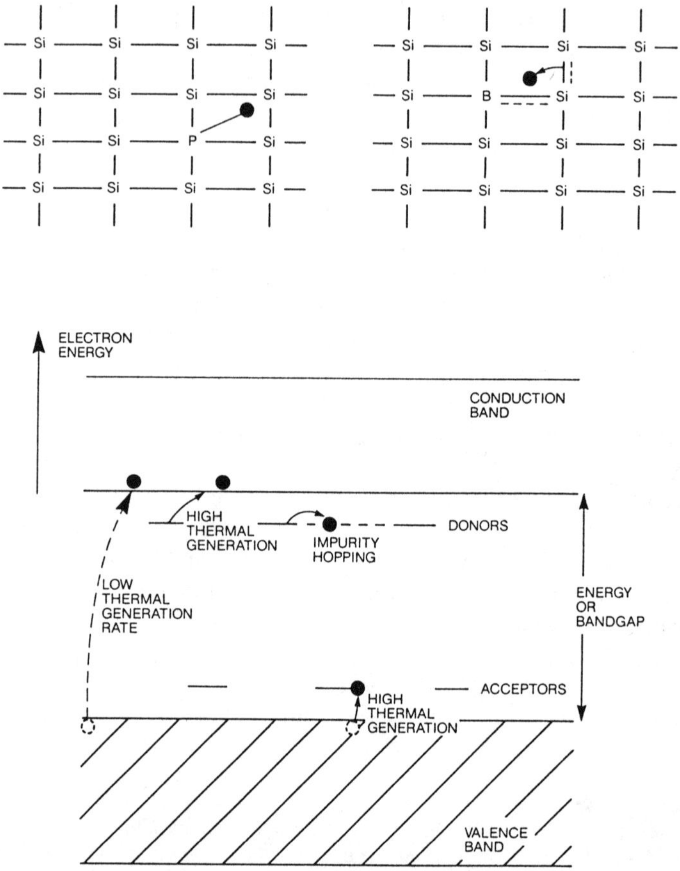

In a two-dimensional representation of the silicon crystal lattice, each silicon atom covalently bonds to four others. A dopant phosphorus atom, which has five electrons in its outermost shell, has an extra unbonded electron left after bonding with the four neighboring silicon atoms. The binding energy of this excess electron is only about 0.1 eV and is called a donor state. If the dopant molecule is boron, which has only three electrons available for bonding, then the unsatisfied bond, called an acceptor state, can accept an electron from a neighboring bond. This also requires only 0.1 eV to occur and leaves a "hole" (dashed line) which moves as though it were a distinct particle with a positive charge. Thermal generation of free electrons and holes from these donor and acceptor states is much more probable than that due to intrinsic excitation directly across the total bandgap of silicon which requires 1.0 eV. Dopants, introduced in controlled amounts, can therefore be used to dominate the electrical conduction. If the dopant concentration is sufficiently high, then the probability grows that an electron in one donor or acceptor state can move to another. This so-called impurity conduction provides another mechanism of electrical conduction in addition to that involving carriers moving in bands. Such extrinsic semiconductors are the basis of solid-state electronics which have enabled microprocessors and modern desktop computers.

deviate from these idealized situations and the electrical properties of semiconductors, in particular, are profoundly affected by imperfections or impurities. As we have seen silicon has four electrons available for bonding. However there are atoms such as phosphorus and boron which have a different number of valence electrons in their outermost occupied shell. First consider the case of phosphorus which can occupy the position of a silicon atom in a silicon crystal and being pentavalent has 5 valence electrons. After bonding with the four neighboring silicon atoms, a phosphorus atom will have one unbonded electron. It turns out that the energy with which this excess electron is bound to the phosphorus atom is substantially smaller than the band gap of silicon. In fact, the binding energy of this electron can be estimated by a simple calculation, since it can be thought of as being bound by a single positive charge on the phosphorus atom. In this sense the system resembles a hydrogen atom except that the electron orbit encloses some silicon atoms which screen the bound electron and reduce its binding energy from the value in the hydrogen atom (13.6 eV) by an amount inversely proportional to the square of the dielectric constant. Since the dielectric constant of silicon is 11.5, the donor energy level should lie about 0.1 eV below the conduction band, so this electron is rather easily thermally excited into the silicon conduction band where it can conduct electricity. Because of this relative ease of donating an electron, phosphorus is called an n-type donor (n for negative charge), and relatively small concentrations of phosphorus doped into silicon can dominate its conductivity. In fact the energetic positions of the states associated with such dopants are determined by experimentally measuring the temperature dependence of the resistivity of doped silicon crystals. This type of state within the forbidden gap is termed a localized state because a carrier which occupies such a state is normally unable to move, except by being thermally excited into the band. An exception to this is if the level of doping is very high. Since the dopant atoms are then physically closer, it becomes possible for the probability to hop from one dopant state to a nearby unoccupied dopant level to be larger than the odds of it being thermally kicked into the conduction band. This type of transport is called impurity hopping and can also occur between localized levels caused not by dopants or impurities but by defects.

In contrast a boron atom can only bond to three neighboring silicon atoms since being trivalent it only has three valence electrons. Consequently an electron deficient bond exists and an electron thermally excited from the silicon valence band is readily accepted and captured into this vacant level; boron doping thus creates an acceptor state. Similar considerations to those previously discussed lead to an estimate of about 0.1 eV for the boron acceptor level above the valence band.[30,36] The electron accepted by the boton comes from the filled silicon valence band. In doing so a "hole" is created in the otherwise filled valence band. The application of an electric field in a silicon crystal which has been doped with boron (or other trivalent impurities) allows electrons in the valence band to acquire energy and move to fill the hole. Such conduction involving the motion of an electron vacancy in the valence band, can be equivalently described as the motion of a positively charged entity called a hole and is termed p-type conduction. Related kinds of donor and acceptor states can be associated not only with other impurities but defects or deviations from the perfect periodicity of the silicon crystal structure. Semiconductors whose electrical conduction is controlled by impurities or defects are called extrinisic.[35,36] Crucial to the technological utility of these phenomena and the creation of the electronic semiconductor industry was the controlled production of precise regions of n and p type conductivity in silicon crystals by the deliberate introduction of various elements as impurities.[35] Doping levels of a few parts per million are used for modern day semiconductor technology so that an ability to grow silicon crystals with exceptionally low levels of background impurities and a high degree of perfection is mandatory.[37] Given this perspective, the source of confusion and ambiguity that early experimenters had to endure explains the operating room-like facilities of today's silicon foundries.

With the recognition that two types of conduction involving electrons and holes can occur in semiconductors, the question arises as to how one establishes this fact. This is where the Hall effect proved so valuable. As mentioned earlier, when a magnetic field is applied at right angles to the direction of current flow in a solid, a voltage called the Hall voltage is set up in a direction perpendicular to both the current flow and the applied magnetic field. The sign of the Hall voltage depends on, and therefore can identify, the sign of the charge carrier producing the current.[33] Application of magnetic fields to semiconductors prior to the advent of the band theory of solids had yielded mystifying and inexplicable results where both negative and positive signs of the Hall voltage could be observed. With the insight of the band theory, the Hall effect was ultimately used to confirm that conduction in metals is due to electrons, identify conduction in nonmetallic solids as being electronic rather than ionic as some had suggested, and finally verify the concept of hole conduction.

Having discussed the basic tenets of the band theory of solids, it is now possible to discuss in a more fundamental way the phenomenon of photoconductivity which is so important in xerography. After its original observation in selenium by Smith in 1873, subsequent progress was slow for several reasons. Initially, the nature of photoconductivity was unclear without prior knowledge of the existence of the electron or the development of the band theory of solids. Indeed one of the earlier interpretations of photoeffects in selenium involved light-induced structural changes, and it was not until 1925 that Lenz, using the Hall effect, established that electronic carriers were involved.[30] In addition, just as in the case of electrical

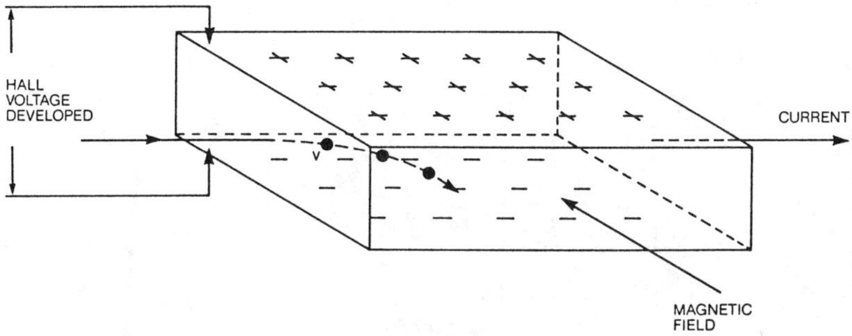

The Hall effect depends on the fact that a freely moving charged particle can be deflected by a magnetic field applied perpendicular to its direction of motion. The deflection of electrons, which occurs in the downward direction in the illustration, produces a voltage known as the Hall voltage between the top and bottom surfaces as indicated. For the same configuration, a Hall voltage of the opposite polarity would be developed if the carrier dominating the conductivity were holes as would be the case for p-type, extrinsic boron-doped silicon. The band theory of solids therefore explains the mysterious observation of positive and negative Hall voltages made twenty years before the discovery of the electron! Since the degree of deflection by a magnetic field of a moving charged particle is dependent on its velocity, the Hall effect can therefore be used to measure the mobility of free carriers.

conductivity in semiconductors, the ability to assemble a coherent set of observations and principles was plagued by the quality and purity of the experimental samples. Thus in contrast to materials available to today's experimentalists, the early investigators were forced to work either with natural crystals of dubious and certainly uncontrolled purity, or powders with a plethora of complicating factors such as interparticle interfaces.

The first systematic and sustained study of photoconductivity was by Pohl and his group in Göttingen in the 1920s.[38] They focussed their efforts on single alkali halide crystals such as sodium chloride and potassium chloride and developed techniques for growing them in the laboratory. They also did some limited studies on the form of selenium called monoclinic. As we have discussed, the alkali halides are almost purely ionic in that the individual sites in the crystal lattice are occupied by either positive sodium ions (Na^+) or negative chlorine ions (Cl^-). At or above room temperature, the resistivity of these solids is controlled by ionic motion. Under an applied field, the positive and negative ions move to the cathode and anode respectively by the thermally activated movement of the ions into unoccupied defect sites in the crystal. Purely electronic conduction can only be observed at temperatures below room temperature where this activated ionic motion is frozen out. In this way they established a number

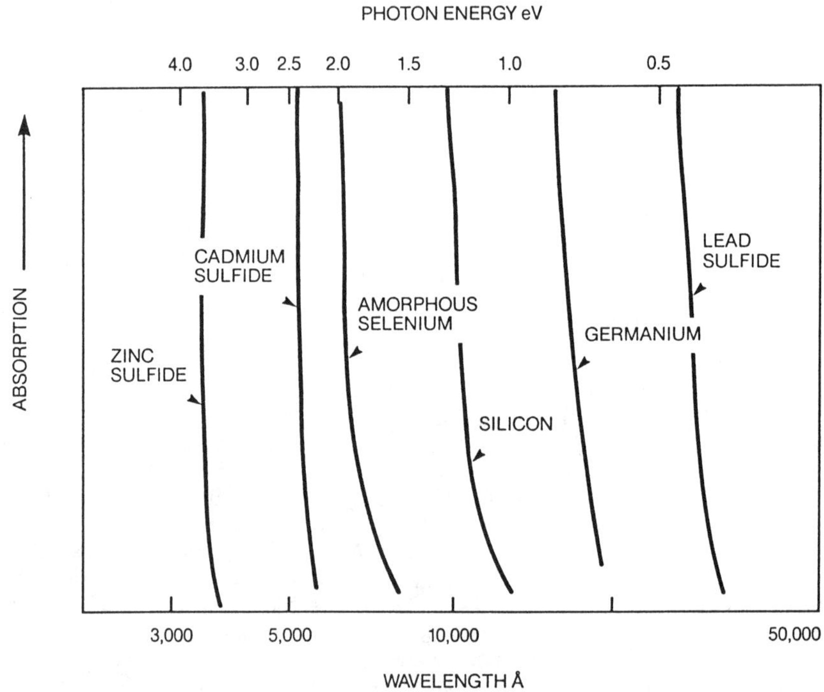

When the energy of incident photons equals or exceeds the bandgap between the valence and conduction band, strong absorption of the light occurs. This threshold is called the absorption edge and accounts for certain solids being colored while colorless solids such as common salt have an absorption edge in the ultraviolet part of the electromagnetic spectrum. Such light absorption is associated with the production of free electrons and holes in the conduction and valence bands and defines the onset of photoconductivity which is the heart of the xerographic process. For copying, photosensitivity to visible light is desirable so that materials with bandgaps of 1.65 eV or smaller are in principle desirable. In practice, if the bandgap is much smaller than this, thermal generation of free carriers across the gap, which occurs even in the dark, results in too low a dark resistivity and the corona charge on the photoreceptor is lost too rapidly for it to be useful.

of basic experimental factors. The photoconductivity was shown to depend on the wavelength or photon energy of the light and that its magnitude was affected by the addition of impurities. Gradually a picture emerged in which electrons usually bound to the constituent atoms could be liberated by light excitation. The free carriers could then move in an applied field for a time before they became immobilized in electronic states called traps.

All of these observations can be accounted for in the context of the band theory.[39] Just as electrons can be thermally excited across the bandgap, they can be transferred by any source with energies equal or greater than the bandgap. The energy can be provided by atomic particles such as α or β-particles or electromagnetic waves, including as x or γ-rays. This book will limit the discussion to the excitation of electrons by the absorption of light photons which is the basis of our visual world and xerography. Associated with photoconductivity is one of the striking visual properties of solids, namely, the existence of an absorption edge. Thus solids do not normally absorb light until the photon energy exceeds the value equal to the bandgap, although in the case of metals their aesthetically striking visible optical properties are associated with the very strong reflectivity in the visible spectrum due to light excitation of the free electron gas within the conduction band. If the absorption edge lies in the visible part of the spectrum, the solid will have a characteristic color. In the case of the alkali halides, the bandgaps are several electron volts wide and their absorption edges lie in the ultraviolet, which accounts for their colorlessness. The existence of an absorption edge leads naturally to the expectation of a threshold in the wavelength of light capable of producing photoconductivity. On exposing a solid, previously kept in the dark, to bandgap light electrons are promoted to empty states in the conduction band leaving corresponding holes in the valence band. These are referred to as photogenerated electron-hole pairs. These carriers can now move in an applied field to produce a photocurrent. Although the electrons and holes move in opposite directions their contributions to this current are additive since they are oppositely charged. The number of photogenerated carriers created per absorbed photon, called the photogeneration efficiency, is an important figure of merit for a photoconductor and is strongly influenced by the purity or perfection of the material. Although constant illumination is possible, the technological importance of photoconductivity relates more to the *changes* that can be produced. As a result, it is the transient aspects of photoconductivity with which this work will be most concerned. Photoconductivity is a nonequilibrium phenomenon, since the normal equilibrium state is when the solid is unilluminated. Consequently the photogenerated carriers may find some lower energy states localized in the bandgap, aptly called traps, which arise from either impurities or defects in the crystal lattice. The average time that carriers remain free before such a trapping event takes place is called a trapping lifetime. Intuitively one can see that the higher the density of such traps the shorter the trapping lifetime will be, since motion in the conduction band represents free carrier motion through the solid. During such spatial motion, the carrier can encounter a position in the lattice occupied by an impurity of where a defect exists; the probability of such an event therefore increases with the density of such perturbations.

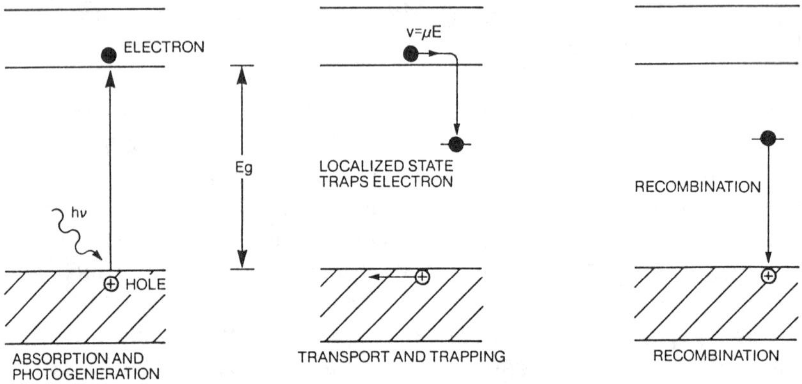

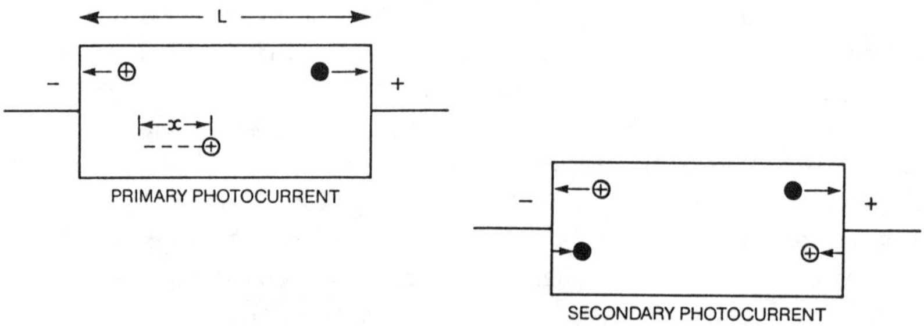

Free carriers produced by absorption of photons with energies greater than the bandgap move with a velocity equal to μE. Localized states, associated with impurities or defects and appropriately called traps, lie within the bandgap and depending on their density can capture free carriers; the time carriers remain free is termed their lifetime, τ. Within a trap, a carrier is immobilized and can only move by thermal reexcitation back into the valence or conduction band, although if the trap density is sufficiently high then impurity hopping may occur. A trapped carrier may attract and mutually annihilate or recombine with an oppositely charged free carrier in a recombination lifetime, τ_R. Such recombination processes are the mechanism by which photoconductivity decays to its equilibrium dark value when light exposure is terminated. The concepts of primary and secondary photocurrents differ depending on whether a photogenerated carrier is replaced after it has traversed the photoconductor sample thickness L. In primary photocurrents, as is the case for xerography, no replenishment takes place and the maximum number of carriers that flow cannot exceed the number created with light, so that only unity gain is possible. Gains less than unity will occur if the ratio of the distance a carrier moves, $x = \mu \tau E$, is less than the thickness of the photoconductor, L. In secondary photocurrents, when one carrier exits at an electrode another can enter at the other electrode and depending on how many passages can be made through the sample in a carrier lifetime, more charge flows than was photogenerated and photoconductive gain can occur. Special electrode materials are usually required to achieve this condition, which is widely employed in sensor devices such as scanners.

While occupying such localized traps, the carriers cannot move and contribute to the photocurrent. Such trapped carriers may also, via the coulomb force, attract and mutually annihilate excess carriers of the opposite sign which pass nearby. Such a process called recombination is the mechanism by which a solid returns to its quiescent dark state after termination of light exposure.[40] For the recombination event one can define a recombination lifetime. These concepts will be developed in more detail later.

Gudden and Pohl distinguished two types of photocurrents, primary and secondary. In primary photocurrents, the relative contributions of the free electrons and holes to the total photocurrent are related to the distance they move in the applied field before being trapped.[38,39] The photocurrent is proportional to the applied field at low fields but saturates at a sufficiently high field. This situation corresponds to the case where the sum of the distances moved by the two types of carriers is equal to the maximum, which is the photoconductor thickness. At this point the photocurrent per absorbed photon is the photogeneration efficiency, which is at most unity since one photon creates one carrier. Since secondary photocurrents are not involved in xerography, they will not be discussed further except to say that secondary photocurrents involve carrier replenishment at the metal contacts or electrodes by which electric fields are applied to the solid. Under these conditions, as a photogenerated electron exits at the anode it is replaced by an electron at the cathode. Since one absorbed photon can result in the flow of thousands of carriers within a carrier lifetime, amplification occurs and photoconductive gains much larger than unity can be obtained. Such secondary photoeffect devices are used to make highly sensitive solid-state silicon photodetectors.[37,39,40]

This basically was the state of affairs in the mid-thirties when Chester Carlson turned his attention to the problem of inventing a radically new, better and cheaper way to make copies. Most of the topics that have been discussed played an indispensable role in the invention, development and innovation of xerography. Others such as the development of solid-state electronics based on crystalline silicon, radio and telephonic communications, and computers together with their software initially played a minor or no role at all. However these parallel technological innovations have become increasingly relevant over the past decade enabling improvements, enhancements and extensions of xerography far beyond Carlson's wildest dreams. It is beyond the scope of this narrative to discuss in any detail these technological advances that gained momentum during World War II and continue to this day.[37,41,42] Suffice to say that they culminated in the invention of the transistor and all that was to mean for the embryonic computer and telecommunications industries. As the story unfolds, appropriate aspects of these technologies will be elaborated on, but it is now time to turn to the story of Chester Carlson and his invention.

Chapter 2
Carlson and His Invention: Dry Images

Chester F. Carlson was born on February 8th, 1906, in Seattle, Washington, of Swedish stock. Shortly after his birth, the family moved several times before settling in San Bernardino, California. Even at an early age, he avidly read about Edison and other inventors and delved into experimentation and construction of projects.[43,44] In particular, he was fascinated with graphic arts and also especially liked chemistry in school. One of the parttime jobs Carlson had to take outside school hours, because his father had an incapacitating illness, was working in a printing shop. In lieu of wages, the owner gave him an old printing press which he used to publish a few issues of a magazine for amateur chemists. The experience of setting his own type impressed on Carlson the difficulty of producing hard copy of words, and planted the seed of thinking about better alternatives.

After graduating from high school, he entered Riverside Junior College for three years before transferring to the California Institute of Technology where he was awarded a B.S. degree in physics in 1930. Not surprisingly, his courses covered very little if any of the advances in atomic and quantum physics being made at that time. After graduation Carlson took a position as a research engineer at the Bell Telephone Laboratories in New York. However he soon found the work uninteresting and repetitious and requested a transfer to the patent department which he thought would prove to be a more stimulating environment because of the inherent exposure to new and creative ideas. Although this particular post disappeared in 1933 during the Great Depression, he found a similar position in a patent attorney's office in New York City. After a short time there he joined the patent department of the P.R. Mallory Company where he was to stay until 1945. During this time he became a registered patent attorney and by studying at night acquired a degree from the New York Law School in 1939.

By 1935 Carlson, newly married and struggling to make ends meet, made a conscious decision to revisit the question of copying documents with the objective of making some revolutionary invention. His motivations were both altruistic, in his own words, "to do the world some good," but also, "to do some good for himself." During the course of his daily work, Carlson must have been constantly aware of the need for more and more copies of documents relating to patent applications. At that time, of course, there were only two alternatives. The first was to have photographic copies made. This was very time-consuming, labor-intensive and consequently expensive to use in any routine way. The second, but equally unsatisfactory approach, was to have a typist or graphic artist produce more original copies which in turn required additional, painstaking proofreading for mistakes. Thus the idea was born of a small machine in the office which could make low cost copies of an original document with acceptable quality at a rate of one every few seconds. Carlson quickly decided that such a machine should be based on the direct use of light. Aware that silver halide photography and other light-induced chemical phenomena were exhaustively being pursued in the research laboratories of major corporations however, Carlson studiously chose to identify and investigate other light-related phenomena.

In the course of an extensive literature search, Carlson systematically examined many relevant reports and patents. Years later, in a revealing and forthcoming comment, Carlson said, "Things don't come to mind readily all of a sudden, like pulling things out of the air. You have to get your inspiration from somewhere and usually you get it from reading something else."[44] It is indeed true that over a period of 150 years prior to his invention a number of effects, even devices, had been explored.[45] In most cases these were isolated experiments, however, involving electrostatic phenomena which at best were viewed, at the time, as mere scientific curiosities. In retrospect, an idea, intent or connection is often inaccurately ascribed to an earlier investigator or assigned an importance which perhaps never existed. Nonetheless these pioneering efforts must have stimulated Carlson's imagination as to what might be possible.

As early as 1777 a German physicist discovered electrostatic recording when he produced the first Lichtenberg figures which immortalized his name.[45] These images, produced by scattering dust on a slab of insulator that had been exposed to an electric spark, were nonpermanent and easily destroyed by blowing on them. A decade later Villarsy, using a mixture of red lead and sulfur powders, was able to distinguish between the two polarities of charge using a Lichtenberg figure, since red lead was attracted to the negative charges and the yellow sulfur adhered to positive charges. The first recorded practical use of electrostatic recording was by Ronalds in about 1840. Based on Lichtenberg figures, it involved the movement by clockwork of a stylus contact, connected to a lightning rod, across an insulating resinous surface.[45] The powder-developed image reflected the intensity and nature of the atmospheric electricity. Intermittently throughout

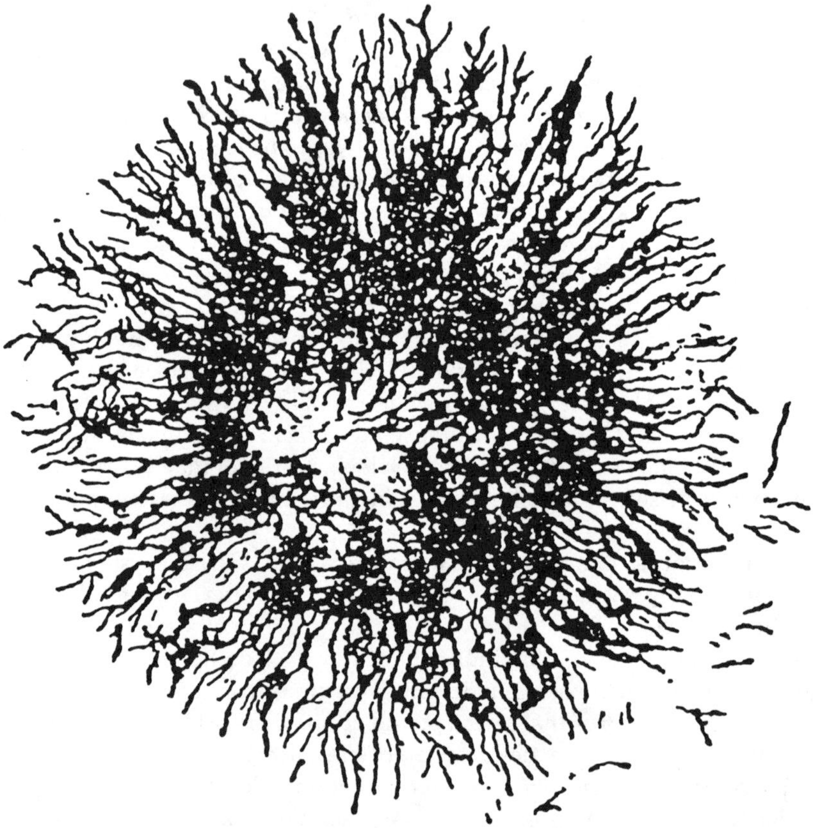

Lichtenberg figures were first produced by George Christoph Litchenberg at Göttingen University in 1777 when he discovered that these dendritic-like patterns could be produced by dusting insulators exposed to an electrical spark.

the late nineteenth and early twentieth centuries, a number of related observations were made in which electric charges were rendered visible. All of them had in common the use or accidental involvement of dust, powder or particulate aerosols including candle or tobacco smoke.

The first serious and sustained efforts at electrostatic reproduction of images were due to Paul Selenyi, a physicist at the Tungstram Laboratories in Budapest, Hungary, in the 1920s and '30s.[46] These processes, which he generically called "electrography," involved the writing of electrostatic images on insulating surfaces with controlled electron or ion beams. Initially the images were drawn on the inside of a cathode ray tube and developed by dusting the outer surface with positively charged lycopodium powder which consists of spores of a genus of moss plants. Other variations followed, including development on an intermediate thin insulating film

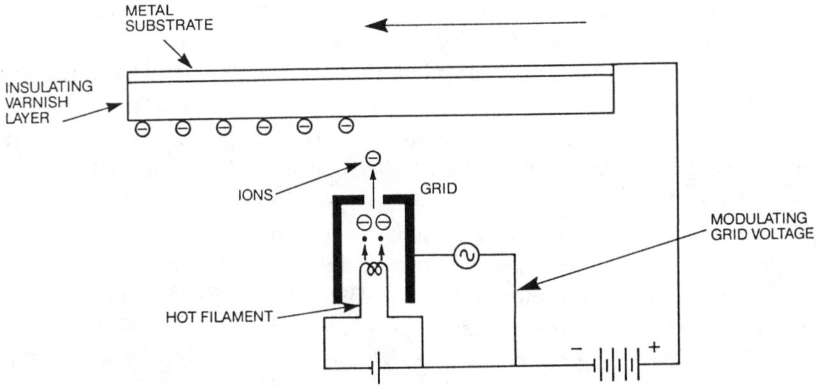

Selenyi developed the first facsimile machine using electrography in 1938. The electrons from the filament formed negative ions by attachment to air molecules. This ion flow, controlled by the application of a modulating voltage to the grid, was drawn onto the insulating layer coated on the positive anode. Using a modulating voltage derived from scanning a photograph with a photosensitive phototube, Selenyi demonstrated by powder development of the ion image that transmitted pictures could be recreated at a distance with electrography.

contacting the outside of the cathode ray tube; this could then be removed together with the developed image. Another approach was to pre-dust the external tube face with a negatively charged powder which was then electrostatically driven off onto a contiguous thin insulating sheet by the negative charge pattern drawn on the inside of the tube by the electron beam. Inevitably, such images were seriously blurred due to the relatively large thickness of the tube wall. To overcome this shortcoming, Selenyi used a Lenard window of cellulose film on a cathode ray tube which was sufficiently thin to allow the transmission of electrons, while strong enough to sustain the necessary vacuum over small areas. The emerging electrons were then attracted to a closely spaced, positively charged metal plate coated with a thin insulating film.

An important extension of electrography by Selenyi was the use of ions rather than electrons. The ions were generated using a hot barium oxide–coated platinum cathode since the electrons liberated from the cathode attached themselves to gas molecules to form negative ions. The cathode, in the form of a wire, was partially surrounded by a cylinder in which there was a thin slit. This cylinder, to which a variable negative voltage was applied, served as the grid or gate as in a vacuum triode tube. With a sufficiently high negative voltage applied, the ions were repelled from the grid and therefore confined. The anode was a metal drum coated with an insulating film and being positively biased attracted the negative ions. The drum could be rotated past the grid aperture and also moved laterally so that effectively it was scanned and after deposition the charged ion pattern could be powder developed. By applying a signal voltage, derived from scanning an image with a photoelectric detector, to the grid, the intensity of the ion beam could be modulated. In 1938, Selenyi reported the transmission of facsimile pictures produced in this way.[46]

Shortly after the discovery of x-rays and radioactivity, a number of attempts were made to use these rays to produce electrostatic images.[45] These were contact images in that the object was laid on a previously charged insulating plate. Subsequent illumination with x-rays, radiation from radium or even ultraviolet light produced charged images that could be powder developed. Such effects were probably produced by the ionization of the air above the plate.

A distinctively different approach was the investigation of the potential value of the photoelectric effect in imaging. The photoelectric effect (sometimes referred to as the external photoeffect) involves the photoejection of electrons out of a solid on its exposure to light and was important for the role it played as an experimental test of the quantum theory. Both Selenyi and Carlson worked on and considered this approach and Carlson filed a patent on September 8, 1938.[47] There were manifold problems, however, with this approach. The range or distance that electrons move before coming to a rest in air is such that an extremely close spacing between the photoemissive surface and the anode imaging plate was necessary. Moreover, the efficiency of this process is intrinsically very low since many electrons are backscattered into the solid. Although emissive surfaces can be produced using materials sensitive to visible light, these are often unstable in air and uniform emissive properties are difficult to achieve over large areas.

The alternative approach was to use the phenomenon of photoconductivity.[39,40] This can be thought of as an internal photoeffect, although it is never referred to as such. A number of essentially unsuccessful attempts had been made by others to use photoconductivity to control electrochemical changes in specially treated paper. These experiences caused Carlson to conclude that, apart from chemical instability issues, the current requirements of these chemically-based phenomena were incompatible with the magnitude of photocurrents that could normally be produced. This led him to experiment with the formation of electrostatic images on photoconducting insulating layers.[44] Much of his early experimental work with a number of possible photoconductive materials was done in the kitchen of his apartment in Jackson Heights, New York. In the fall of 1938, Carlson hired a recently arrived refugee engineer from Germany, Otto Kornei. He also rented a back room on the second floor of a building in Astoria, Long Island, and within weeks the reduction to practice was made. The event is best described in Carlson's own words[43]:

> October 22, 1938, was an historic occasion. I went to the lab that day and Otto (Kornei) had a freshly-prepared sulfur coating on a zinc plate. Otto took a glass microscope slide and printed on it in India ink the notation, 10.−22.−38 Astoria. We pulled down the shade to make the room as dark as possible, then he rubbed the sulfur surface vigorously with a handkerchief to apply an electrostatic charge, laid the slide on the surface (of the sulfur) and placed the combination under a bright incandescent lamp for a few seconds. The slide was then removed and lycopodium powder was sprinkled on the sulfur surface. By gently blowing on the surface all the loose powder was removed and there was left on the surface a near-perfect duplicate in powder of the notation which had been printed on the glass slide. Both of us repeated the experiment several times to convince ourselves it was true, then we made some permanent copies by transferring the powder images to wax paper and heating the sheets to melt the wax. Then we went out to lunch and to celebrate.

Kornei stayed with Carlson until April 1939 during which time a number of advances were made. Amongst these was the fabrication of more photosensitive films by the sublimation and sintering of an organic compound, anthracene, onto zinc plates. On April 4, 1939, Carlson filed the first seminal patent on what he called electrophotography.[48] The initial flush of success was, however, followed by four years of frustration. Working alone once again, Carlson designed and built an automatic copying machine which he filed for patenting in November 1940.[49] Although the progress was slow, due to the shortages and priorities associated with World War II, Carlson captured the essential elements of the xerographic process with this prototype automatic copying machine.

It is necessary at this point to describe the basic features of the integrated xerographic process even though this means anticipating some of the necessary breakthroughs to come, for other inventions and improvements ultimately proved necessary to translate the original invention into a technological and commercial success.[50,51] Many of these basic elements of the xerographic process will be expanded on in later chapters. Figuratively and literally the heart of a xerographic machine is the photoreceptor which consists of a thin film, 20 to 50 microns thick, coated onto a grounded aluminum drum or metallized belt. The film material must have a high resistivity in the dark ($\sim 10^{12}\ \Omega$ cm) and yet be photosensitive on exposure to visible light. The photoreceptor drum is mechanically rotated with respect to the other necessary subsystems. The first of these, a

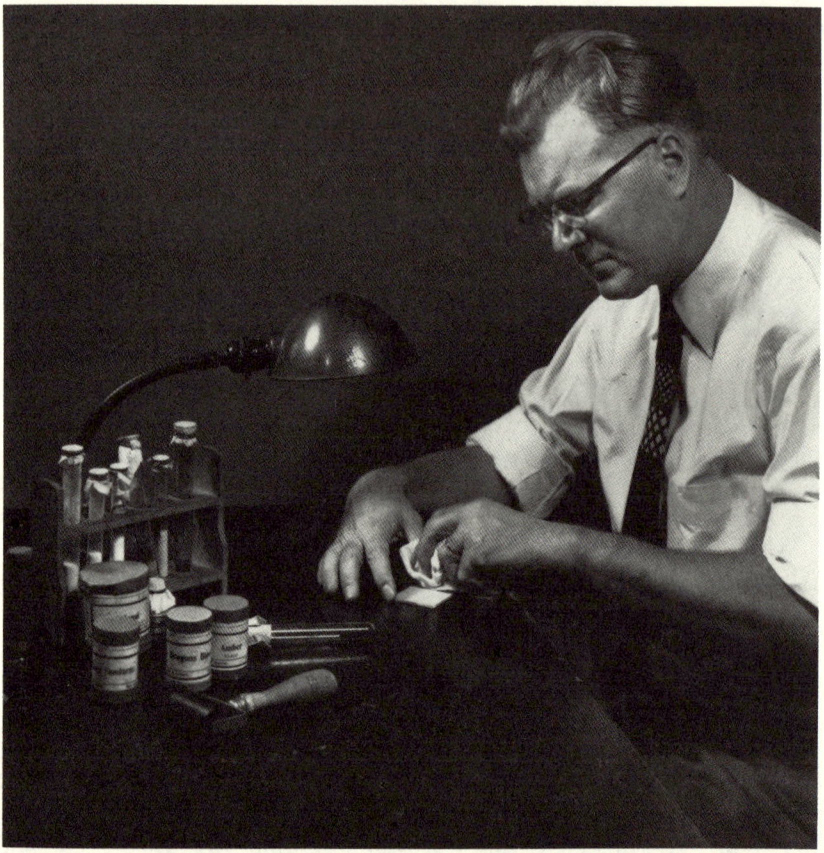

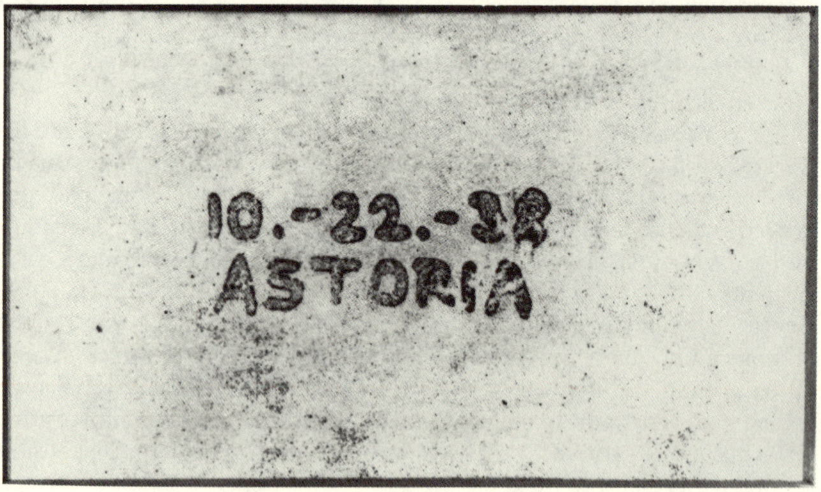
10.-22.-38
ASTORIA

corotron, is a charging device which produces either positive or negatively charged air ions depending on the polarity of a high voltage applied to a wire. These ions are accelerated by the applied voltage and deposit on the surface of the photoreceptor. This results in the photoreceptor film charging as a parallel plate capacitor. As shown by Faraday, the voltage across the film is proportional to the number of ions deposited per unit surface area and to the inverse of the material's dielectric constant. Typically voltages of a few hundred volts are used to provide the necessary development fields. This surface potential, divided by the film thickness, determines the applied electric field within the photoreceptor. The charged area of the photoreceptor then rotates, with minimal loss in the initial voltage because of its high resistivity in the dark, into the exposure position. In this position an optical image, produced by scanning the document to be copied with a slit light source, is focussed by lenses onto the photoreceptor surface. Where light reflected from white areas of the document strikes the photoreceptor, photogeneration of free electrons and holes occurs within the absorption depth of the light, which is typically much less than the film thickness. The dark characters on the original document reflect less light and proportionally fewer photocarriers are created. Assuming the initial voltage is positive, photogenerated holes move across the photoreceptor film and electrons move the shorter distance back to the top surface under the influence of the electric field within the photoreceptor. This causes the initial voltage to selectively discharge to a degree depending on the amount of light locally absorbed by the photoreceptor.

Leaving the exposure zone, the photoreceptor now carries an invisible distribution of surface charge or voltage which replicates the document image. This latent electrostatic image is then developed by cascading charged dry ink or toner, consisting of polymer particles containing carbon black pigment, over the photoreceptor surface. The toner charge which is produced by controlled triboelectric charging is opposite in sign, negative in this example, to the original charge deposited on the photoreceptor. The negatively charged toner particles are attracted by their mutual coulomb electrostatic force with the remaining positive surface charges of the latent electrostatic image and become attached to the photoreceptor surface. After leaving the development zone the original document charge image is

Opposite (Top): *Chester Carlson, reenacting the production of the first xerographic image, is charging the thin sulfur film by rubbing its surface with a handkerchief. The lamp used for the light exposure and various bottles of dry powder used for image development are also visible.* (Photograph, Courtesy Xerox Corporation) Opposite (Bottom): *A photograph of the first xerographic print, which Carlson saved. The darker edges of the characters compared with the grayer inner solid areas are due to the dominance of fringe-field development.* (Photograph, Courtesy Xerox Corporation)

Oct. 6, 1942. C. F. CARLSON **2,297,691**

ELECTROPHOTOGRAPHY

Filed April 4, 1939

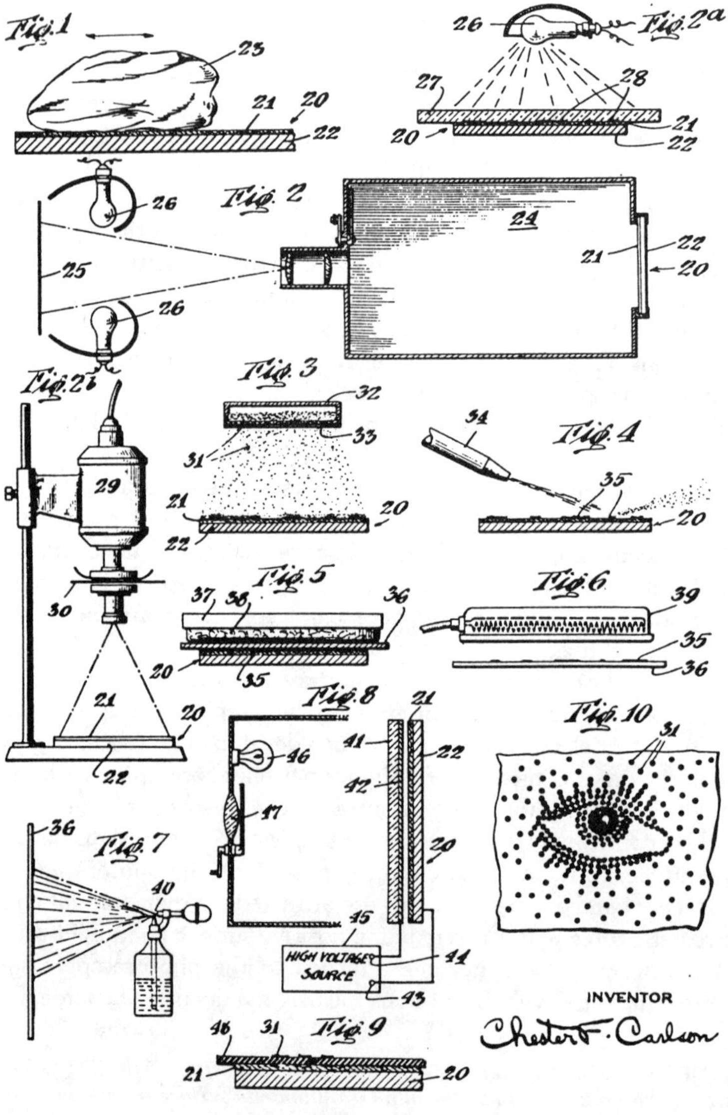

The first page of the seminal patent on electrophotography awarded to Chester Carlson; although the patent was filed in 1939 it did not issue until 1942.

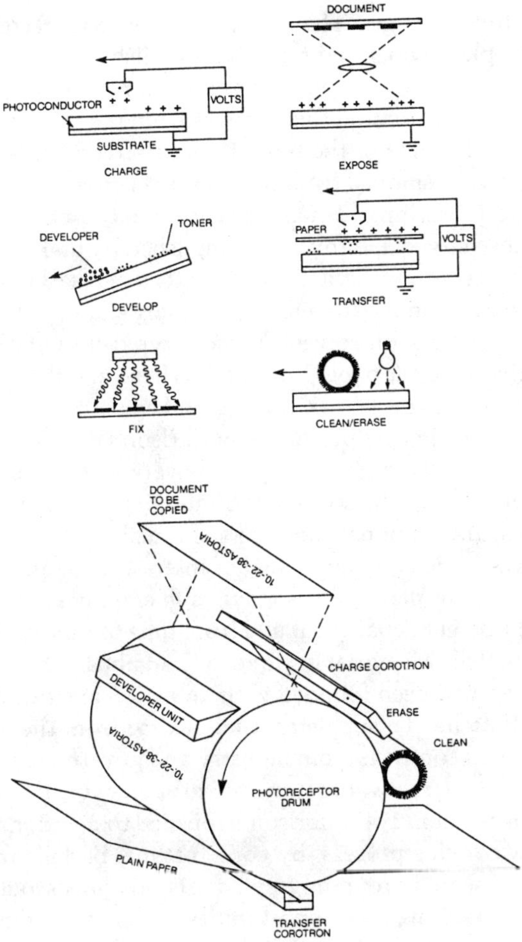

The basic process steps used in most xerographic products include: the charging of a thin photosensitive insulator, the photoreceptor, with ions generated with a corotron device; exposure of the charged photoreceptor to an optical image during which photodischarge of the photoreceptor proportional to the light intensity in the optical image occurs; the selective discharge produces a latent electrostatic image which is rendered visible or developed with triboelectrically charged toner or dry ink; the toner is next transferred to paper by charging paper to the appropriate polarity to attract the toner from the photoreceptor surface; once on the paper the toner is fixed or fused to the paper by softening the polymer-based toner either with heat or solvent; finally the residual toner is removed from the photoreceptor by mechanical wiping and the residual latent charge image destroyed by exposing the photoreceptor to uniform illumination over its entire surface. If the photoreceptor is in the form of a drum or a belt, then these steps are compatible with a rotary configuration such as indicated and can occur cyclically in modern high-speed machines at speeds of 120 documents per minute.

now a visible toner image. This must next be transferred from the photoreceptor to plain paper and this is accomplished by using a transfer corotron. This charges the paper with ions of the same polarity as those used to charge the photoreceptor and toner moves from the photoreceptor onto the paper. Since only some of the toner is transferred in this step, the remainder must next be removed from the photoreceptor surface by, for example, mechanical brushing. Finally any residual latent-image surface charge must be discharged and this is accomplished by uniformly exposing the entire photoreceptor to light from an intense erase lamp. After the transfer step, the paper moves to a fixing or fusing station where the toner image is softened either with solvent or more typically with heat. In this way the toner adheres to the paper, the image is fixed and the copy is ready. At this point the process cycle is completed and can automatically restart.

Unfortunately the images produced by Carlson's original embodiment were not of particularly high quality and were marred by unwanted powder in the white image background. So despite contacting nearly two dozen companies, the National Inventor's Council and the Army Signal Corps, no one was sufficiently impressed or interested to take up Carlson's invention.[52] During this period Carlson was still employed by the Mallory Company in their patent department and from time to time business affairs took him to the Battelle Memorial Institute in Columbus, Ohio. The Battelle Memorial Institute had been founded with an endowment of $3.5 million left by Gordon Battelle. To supplement its income from the endowment, the Institute solicited contracts from industry and government agencies. In return, the research results became the property of the contracting organization. On a visit in 1944, Carlson happened to mention that he held patents on a new copying process. By good fortune, Battelle was thinking of establishing a research group in graphic arts and was scouting around for interesting and exciting new opportunities in that field. Despite the all too evident problems with Carlson's invention, Battelle felt substantial contributions could be made with their expertise in metallurgy, chemistry and physics.[43]

As a result the Battelle Development Corporation, a wholly owned subsidiary of the Institute, acquired exclusive rights to Carlson's patents and agreed to give him a substantial share of any proceeds from subsequent developments. For its part, Battelle agreed to commit three thousand dollars to further develop Carlson's invention which he called electrophotography. Carlson was to recieve 25 percent of all profits or royalties accruing from his invention, or, if he reimbursed Battelle for their investment within five years, his royalty share would increase to 40 percent. In the fall of 1944, a Battelle group headed by Dr. R.M. Schaffert began its work. Its first approach was an investigation of photoconductive materials and the conditions for obtaining a sharp electrostatic image. This was to be followed by a study of the best means for rendering this latent electrostatic image visible, consistent with an ability to transfer it to paper. The possibility of using a liquid developer ink as well as dry powder was an early option considered. In the meantime Battelle encouraged Carlson to try and interest some company in electrophotography as a means of increasing the development effort. As was the case earlier, despite many overtures to major U.S. corporations, no one wanted any part of it.

In 1944 an article on electrophotography, written by Nicholas Langer and published in a magazine called Radio News, came to the attention of a small photographic paper manufacturer, the Haloid Company in Rochester, New York.[53] Haloid, founded the year of Carlson's birth, produced and marketed sensitized photographic paper; the name Haloid was derived from the halogen-salt sensitizers used.[54] As early as 1912, the company employed a large force of salesmen and demonstrators and established a direct sales approach. In 1935 the firm acquired the Rectigraph Company which was the original manufacturer of photocopying machines, first introduced into the United States in 1910. As a result Haloid had a complete line of photocopying products using wet chemistry and including machines, supplies and services. After World War II materials and labor costs were growing faster than the prices that could be charged for the sensitized paper and the company was actively searching for new products. As a result of reading the article, the president of Haloid, Joseph Wilson, and his research director, John Dessauer, visited Battelle to look at the new invention that might enlarge the business opportunities of Haloid.[54] They were sufficiently interested for Haloid to commission a market research firm to see if a market for a new copying device existed. Inevitably, of course, the answers were ambiguous because the device was hypothetical and it was impossible to answer the most basic questions as to advantages, cost, size and speed. Still Haloid's interest in the process continued and in April of 1946 an agreement was signed by which Haloid was to support the Battelle research at a level of $25,000 a year, beginning in 1947. This was not a inconsiderable risk considering that in 1947 the company had a net income of only $138,000 on net sales of approximately $7 million. Essentially all the research on Carlson's invention in the period 1944 through 1948 was therefore carried out at Battelle, initially through its own funds but, starting in 1946, with the support of Haloid.

During this four year period, significant and vitally important advances were made by Schaffert's group that enabled the ultimate commercialization of electrophotography.[55] These included improvements in ion charging processes, photoconductor materials and their fabrication, developer materials and processes and electrostatic transfer.

As noted before, the photoconductive plates that Carlson used to reduce his invention to practice were sulfur and anthracene films, typically a few tens of microns thick.[45] Using the ASA exposure index as a figure of merit, the light sensitivity of sulfur corresponded to a black and white material with an ASA index of 0.002. Anthracene was marginally better with an index of 0.008. Moreover the spectral sensitivity of these two materials was limited to the near-ultraviolet or blue part of the spectrum. Battelle started off by trying to improve the sensitivity of sulfur by introducing small amounts of selenium. Even a few percent made a marked improvement but the improvement was progressive as more and more selenium was added. Eventually it was found that the optimum composition was 100 percent selenium and the amorphous selenium photoreceptor with the required visible light sensitivity was born.[56,57] During this same time period it was also discovered that selenium photoreceptors could also be "photodischarged" by x-rays giving rise to xeroradiography.[50]

The first step in Carlson's process is that of charging the photoconductive layer. In Carlson's initial experiment, this was accomplished using a handkerchief. Today, charging is carried out by spraying the photoconductor with positive or negative ions formed by a corona discharge. Such a discharge is achieved by applying several thousand volts to fine wires. The intense electric fields around the wires cause the air molecules to ionize and charged ions, whose polarity depends on that of the high voltage, are driven onto the photoconductor surface. Two types of corotrons, as these devices are called, were developed.[50,51] The single wire shielded corotron is a relatively simple device which operates in the manner just described and is easy to maintain and clean. A second type known as a "scorotron" is more complicated and is analogous to a vacuum triode tube. A high potential of approximately 7000 volts is applied to the central wire, while an equipotential voltage of approximately a few hundred volts is applied to screen or grid wires. If the potential of the photoconductor surface tries to rise above the screen potential, charging ceases since the ions are no longer accelerated. Thus the surface potential of the photoreceptor is selflimiting and overcharging is prevented.

Carlson employed the simple development method of dusting, first used by Villarsy and others, to develop the latent electrostatic image. Thus the charged and exposed photoconductor film was dusted by the use of powders sifted through screens. Some of the particles were triboelectrically charged by contact with each other or the container, others remained uncharged or may even have had opposite sign charges. Thus although charged particles adhered to the image areas, usually there was powder deposited over the entire film which had to be blown off by an air jet. In the early days, before the research began in earnest at Battelle, this dark background was the Achilles heel of Carlson's invention.

Another major contribution by the Battelle laboratories was cascade development. The new developer consisted of two components and it is

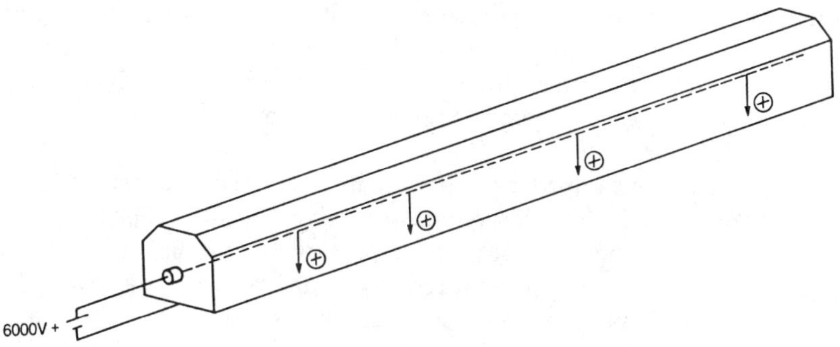

The basic features of a corotron are a thin wire centrally strung within, but insulated from, a metal enclosure which is open on the downward face. Several thousand volts of appropriate polarity applied to the wire produce electrical breakdown which ionizes the surrounding air. The resultant ions are driven out through the aperture. More sophisticated devices such as the scorotron involve control of the ion current to avoid overcharging of photoreceptors.

often called two-component development.[50,51] The first, called toner, consisted of small polymer spheres (approximately 10 microns in diameter) within which even finer carbon black pigment was dispersed. The second component, the carrier, consisted of rather large, more granular material with particle diameters of several hundred microns. The two components together comprised the developer which was cascaded over the photoconductor surface. The carrier's mass controls the cascade and also charges the toner to the correct polarity by triboelectrification. As a toner-covered carrier particle tumbles over an electrical discontinuity in an image area, the vertical component of the fringe field captures the toner particle by overcoming the toner-carrier bond. The two-component system dramatically reduced the amount of toner deposition in the discharged background areas.

Even after the latent electrostatic image had been developed, a critical step was the transfer of this tone-developed image to the required medium before fixing it to render it permanent. Ideally such a transfer should occur without any loss of sharpness and with as high an efficiency as possible. Obviously the original techniques of manual pressing in a soft matrix was totally inadequate for practical use. Schaffert conceived a solution of using electrostatic transfer. The piece of paper or clear plastic sheet, for making transparencies, was laid onto the photoconductive layer over the developed image. The rear of this medium, e.g., paper, was next charged with a transfer corotron using the same polarity ions as used in the original charging of the photoconductive layer. In this way, the negatively charged toner was drawn from the photoconductor onto the paper which finally was peeled away. The image on the paper was then fixed using solvent to soften the toner.

As these various improvements were made, the commercialization of Carlson's invention inched closer and closer to reality. This raised the issue of disclosing the technology publicly to protect the significant Haloid investment, avoid being preempted by some other unknown dry photoprocess and, perhaps more significantly, to attract the additional, substantial financing that would be required to develop a product. One of the first issues was to invent an alternative name to electrophotography which was deemed not to conjure up the required aura of a new, unique invention. It was finally decided that the Greek words for dry, "xeros" and writing or drawing, "graphein" should be combined to give the name xerography to the invention and XeroX (sic) as a tradename.[54]

The public announcement of the new technology was made at the meeting of the Optical Society of America on October 22, 1948, the tenth anniversary of the first xerographic copy. The invention of the transistor was announced on June 30th the same year; although neither was a product at that time. In this same year, Battelle was asked by the U.S. Army Signal Corps to bid on a contract to develop a dry photographic process. Battelle suggested to Haloid that this might be a source of critical funding for the further development of xerography. Eventually the Signal Corps awarded a grant of $120,000 and this proved a major stimulus both to the morale of those involved and the future research efforts.

The first actual product produced was the XeroX Copier Machine, Model A in 1949. It basically consisted of three sub-units, a charging and developing unit, a camera for exposure, and a radiant heat fusing unit. It was a totally manual process involving about a dozen steps before a copy could be produced. This was field tested by loaning units to four large companies who, without exception, found them of little value. Their operation was too complex and copy yields of acceptable quality were highly operator dependent and unpredictable. The obvious solution was to automate the process but this required further substantial investment and investments were unlikely in the face of the experiences with the Model A. Furthermore there were major technical uncertainties in the undemonstrated cyclic stability of the process and the reliable handling of paper within a machine running automatically.

Yet another reprieve arrived, however, when it was realized that the process could be used to make offset lithographic master plates. The process was identical to making single copies except that the image on the xerographic photoconductor was transferred to a specially prepared offset

The nonautomatic XeroX Copier Model A required a series of process steps to produce a copy. A selenium photoreceptor in the form of a 9-inch by 14-inch plate was inserted in the top of the machine in what became a light-tight box. After corona charging and contact exposing, the selenium plate and light box were inserted in the lower, extended assembly which contained the dry developer. After removing a light-shielding panel, the whole assembly was rotated causing the developer to cascade over the photoreceptor and develop the latent electrostatic image. Beyond this stage, the developed selenium plate could be exposed to room light and a sheet of paper was carefully placed onto it. This was then inserted into the central slot seen in the photograph where the toner image was transferred to the paper by charging. Finally, the sheet of paper together with its toner image was warmed in a separate oven unit to melt and fuse the image. (Photograph, Courtesy Xerox Corporation)

A XeroX Copyflo continuous printer that automatically made prints up to 24 inches wide from microfilm originals at a rate of 20 feet per minute. Introduced in 1955, it was the first commercial rotary xerographic product and used a selenium drum photoreceptor. The continuous roll or web of plain paper used is clearly visible. (Photograph, Courtesy Xerox Corporation)

master plate. Such offset plates produce large numbers of copies quickly and at low cost on small lithographic presses. One large user of such processes, the Ford Motor Company agreed to test xerographic photoconductor plates for this purpose. Masters produced by conventional methods required at least an hour to make and cost several dollars each. By contrast, the master made by xerography could be turning out copies within minutes of receiving the original at a cost of 30–40¢ per plate. Thus an important niche had been found in reproduction shops for the embryonic technology, but more importantly, it provided cash, knowledge and experience, all of which were needed to achieve the objective of automating xerography.[54]

The next xerographic product was introduced in 1955. Named the Copyflo printer, it was the first automated xerographic product and enabled the production of copies on a continuous web of ordinary paper from either an original document or microfiche image. By 1955 Haloid's annual sales had reached $21 million of which almost 50 percent came from xerographic products. As recognition of this fact, the name of the company was changed to Haloid-Xerox Inc. The company had also bought from Battelle all rights, title and interest in the original Carlson patents for 50,000 shares of Haloid-Xerox stock which was worth about $4 million at the time.

These transactions had substantial implications for the company's cash flow position which is the amount of money available to a company as a result of profits, plus depreciation and amortization. Cash inflow does not involve depreciation and amortization, but they were deductible for tax purposes. So, by reducing the company's tax liability and its cash outflow, the pool of available cash to finance new projects, facilities etc., was enhanced.[58] In addition, the patents bought from Battelle for $4 million could be amortized over their remaining life at a rate of over $1 million a year. Moreover, the company had decided to lease rather than sell its xerographic products. This was based on a number of considerations. It was felt that a low lease rental would generate more customers for products based on a new largely unproven technology than a high up-front purchase price. The counter argument was that the costs of the product would, in this way, only be recovered over a period of years compared with the immediacy of an outright sale. But it was realized that in the long run substantial benefits could accrue to leasing. First, after the product had amortized its production costs, all future continuing lease fees would be pure profit. Further, only the owner of the machine could depreciate its costs. Thus by leasing, Haloid-Xerox retained legal ownership and the significant tax advantages of depreciation. From 1951 to 1955 the company's cash flow improved from $720,000 to more than $2 million.

In this time frame, Haloid-Xerox, anticipating that its interests would become worldwide, negotiated a partnership in 1956 with the Rank Organization in the United Kingdom to create Rank-Xerox Limited.

Somewhat later, in 1961, a similar arrangement with Fuji-Photo Film Company led to the formation of Fuji-Xerox Corporation.

In 1958 the decision was made to enter the office copying machine market with a xerographic product to be called the Xerox 914, so named because it would make copies on paper up to 9″×14″ in size. Significant research and engineering had to be done even to demonstrate the first model; a coated selenium drum, optical systems, paper feeder, paper transport, fusing and more being needed. Based on the results of the first prototype and manufacturing cost estimates, it became obvious that millions of dollars would have to be expended to bring the product to market. Given the risks involved, several large firms were approached to see if they would manufacture the product for Haloid-Xerox in return for a share of the profits. Market surveys made for these companies, confirming similar ones done for Haloid-Xerox, estimated the maximum number of units that would be sold over the life of the 914 at no more than 3,000. A particularly troublesome issue was the extant market. The 914 was targeted at what might be called the convenience sector of office copying. Such rapid automatic copying already existed. These included electrophotographic products introduced by RCA and which used paper coated with dye sensitized zinc oxide. Such coated papers met market resistance in terms of aesthetics and cost so much that in 1958 the total market was only about $100 million. Moreover, the 914 represented radical departures from the existing copiers and, despite the touted advantages of xerography, these were viewed as highly negative. First, the projected size of the 914 was considered much too bulky and, at 650 lbs. too heavy for the office environment. Second, in terms of a conventional sales approach, was the high cost of an outright sale. The 914 clearly generated no euphoria either in market researchers or other companies. Consequently Haloid-Xerox had to face the necessity of going it alone and establishing its own manufacturing capability or abandoning the whole venture. They chose the former path and, as the saying goes, the rest is history.[54]

As great as Carlson's technological invention and Battelle's improvements were, they were matched by the creative marketing necessary to complete any successful innovation. A major role in the development and implementation of the new marketing approaches was played by C. Peter McColough, a future president of the Xerox Corporation.[54] These approaches had a number of interrelated elements.[58] The central feature was the use of a two-tier pricing structure. This was based on the premise that the ultimate cost to the customer should be a function of the number of copies produced. First, the base price, which included a number of "free" copies, was set sufficiently high to ensure an acceptable return on its investment to Haloid-Xerox. The proviso was that on a per copy basis the base price would be such that the customer would decide that the value of the

The Xerox copier Model 914 was one of the most successful single products in the history of business. By making document copying automatic, simple and affordable it was a truly revolutionary innovation. Introduced in 1959, the last order was taken in 1976 although field service continues to this day. (Photograph, Courtesy Xerox Corporation)

914 exceeded the charge per copy. In addition, the base price, ultimately set at $95, had to be low enough and the contract term short enough to reassure any customer's concerns regarding the investment risk. Meter pricing was then included so that copies in excess of those included in the base price were charged to the customer. Consequently, the higher the utilization of the machine, the higher the revenues and profits. This, in turn, would justify the required establishment and maintenance of a nation-

A modern-day product, the Xerox 1090 Marathon copier, provides a high level of productivity for the user at a speed of 92 copies per minute. It can handle multi-page originals and computer fanfold automatically, has both enlargement and reduction capability and can copy in the duplex mode. A sophisticated information system keeps the user fully informed of the job status. (Photograph, Courtesy Xerox Corporation)

wide sales and service force. The latter was a *sine non qua* for a leasing strategy but in turn also served the additional function of providing the mechanism which facilitated the rapid diffusion of the 914. An ancillary benefit to Haloid-Xerox was, as previously discussed, the depreciation tax benefits accruing to the company by retaining ownership of the machines. As was the case in the technical aspects of this innovation, these marketing innovations did not emerge from nowhere but were almost certainly used because Haloid-Xerox had already experimented with some aspects of these approaches in its earlier history.

The Xerox 914 copier was first shown at the Office Equipment Exhibition in Washington D.C. in September 1959, the New York Business Show in October the same year and it made the cover of *Business Week* on September 19, 1959. It was a fully automatic machine that required no particular skill to operate and no adjustments. Although very basic by modern standards, with no duplex (two-sided copying) or reduction/enlargement, it produced 7 copies per minute on plain paper with a photosensitive element or photoreceptor which was completely reusable. The image was composed of toner which was impervious to degradation either by light or chemicals and ideally suited for archival storage. Its impact was nothing short of phenomenal and it caused a revolution in the copier industry and office alike. In contrast to the market predictions of 3,000 placements over the life of the product, a total of over 200,000 914s were sold, with a peak population of approximately 65,000 units being reached within six years. A monthly copy volume of 50 million in September 1961 climbed to 490 million by March 1966.[58]

The inexorable growth of Xerox's business was slowed neither by economic recession or seasonal fluctuations so that the financial impact was equally dramatic. The revenues of Haloid-Xerox in 1960 were $37 million of which only $3 million were from xerography; including some products such as thermography and diazo. One year later the revenues of the newly-named Xerox Corporation had leapt to $59.5 million and by 1966 revenues from xerography were $428 million.[58] In 1986 the total copier business worldwide for all companies had grown to roughly $20 billion and the annual total number of copies made exceeded 2.5 trillion.[59]

Impressive as these figures are, the evolution and growth of xerography should not be described only in terms of a financial success. Behind the numbers, there exists a story of continued research and advances in materials and phenomena which have led to dramatic improvements in xerographic technology. Likewise, parallel developments in other areas of science continued throughout the years during which Carlson, Schaffert, Wilson and others struggled to make xerography real. As we shall see these apparently unconnected advances ultimately have impacted and enlarged the capability beyond any of Carlson's expectations. It is not hard to believe that Carlson would find today just as exciting and dramatic as that October day in 1938.

Chapter 3
Imaging Materials and Processes: Photons, Electrons and Glasses

Carlson's invention had provided a new way to create copies. However the original embodiment in terms of the materials he used was very far from a practical device. He had identified, at least in principle, that photoconductive, noncrystalline thin films could be used to produce an optical to electrical transducer of the required dimensions and appears to have been intuitively drawn to this approach. With hindsight this choice appears obvious, but had Carlson been totally influenced by the state of knowledge in 1938 he might have been inclined to drop the whole idea. In terms of known photosensitivity, crystals were clearly the best choice, although hardly a practical basis for a low cost way of making copies the size of documents. The best known large-area solids were glasses which are one member of a general class of materials known variously as noncrystalline, disordered or amorphous solids. Amorphous solids are characterized by the fact that, in contrast to crystals, their atoms are distributed randomly rather than on a periodic lattice.[60,61] As a result, x-ray crystallography of amorphous materials which lack well-defined diffracting planes of atoms does not yield the sharp diffraction patterns observed by Von Laue in crystals. On the other hand, freed from the constraint of positional order, amorphous materials can be produced in large areas of arbitrary shape with much less stringent growth conditions. For these reasons, by the 1930s, glass was already a centuries-old established technology, while crystal growth largely had been left to nature. The applications of glass technology were, however, limited to such mundane uses as weather-impervious but light-transmitting windows, household goods or the making of scientific and optical instruments.[62] Despite extensive, albeit indirect, investigations over the years in which glasses had been exposed to all kinds of radiation and electric fields, there was nothing to suggest that such materials were even remotely interesting from the perspective of photoelectronic properties. Indeed glasses were valued for their extremely high resistivity and viewed as quintessential insulators so that the direction Carlson took was

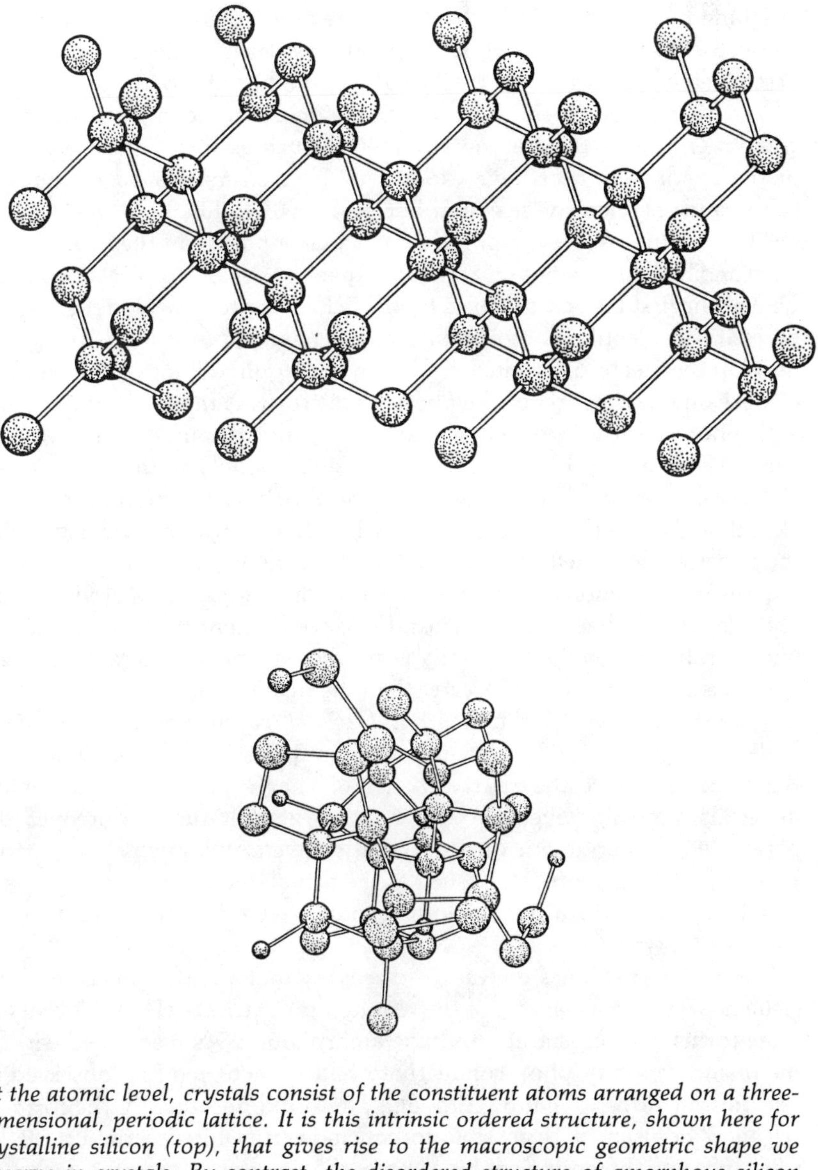

At the atomic level, crystals consist of the constituent atoms arranged on a three-dimensional, periodic lattice. It is this intrinsic ordered structure, shown here for crystalline silicon (top), that gives rise to the macroscopic geometric shape we observe in crystals. By contrast, the disordered structure of amorphous silicon shown (bottom) indicates how in amorphous or disordered materials the atoms are randomly arranged with respect to each other. This causes bond distortions and unsatisfied or dangling bonds; the smaller spheres represent hydrogen atoms which attach to such bonds in the technologically important plasma-deposited amorphous silicon discussed later. Disorder greatly relaxes the constraints on growth form so that amorphous large-area thin films, ideal for image transducers, are readily and cheaply produced.

anything but obvious. Nonetheless, in the best tradition of invention, he chose the route which met his general objectives, independent of any preconceived notions as to what might or might not work.

In his original patent, Carlson had claimed that the required thin films could be made by the melting of materials such as sulfur and anthracene or the condensation of their vapors.[45,48] He also mentioned mixtures of sulfur and the element selenium but even in 1946 his films were at best weakly photosensitive, responded particularly poorly in the visible spectrum and had dimensions much smaller than the size of a typical document. One of the first tasks recognized by the Schaffert group at Battelle was the necessity to identify photoconductive layers of improved sensitivity and develop the means of producing them in the required dimensions. Implicit in Carlson's description of the photoconductor was the requirement for a high photosensitivity which would extend throughout the visible spectrum.[63] This had to be coupled with a high resistivity in the dark so that the initial charging could be preserved until light exposure occurred. It is clear that Carlson also understood that the latent image formation involved the selective, localized photodischarge of a charged capacitor to produce the surface potential difference between discharged and undischarged regions and hence the electrostatic development field.[48] Optimum photodischarge requires a mechanism whereby photogenerated charges move distances equal to or greater than the thickness of the photoconductor layer. It is equally clear that Carlson could not possibly have understood how this might occur or how to specify the necessary material requirements, since the characteristics of charge motion in amorphous materials were not even speculated upon at that time. Moreover the parameters for optimum device operation were unknown. As we now know these depend on details such as the actual development system used and the voltages to which the photoreceptor can be charged without dielectric breakdown.

A major enabling step was the discovery that a particular form of the element selenium has many of the required properties. Selenium occurs in three forms, two crystalline and one amorphous.[63] As mentioned earlier, the phenomenon of photoconductivity had in fact been first observed in selenium in 1873 by Smith. Although Smith did not identify the form of selenium he used, it is possible to conclude from his description of the resistance and dimensions of his samples that its resistivity in the dark was approximately 10^6 (Ω cm).[31] This identifies his material as the form of selenium that goes under a variety of names, "metallic," "gray," or trigonal selenium; one of the two crystalline forms of selenium it has a crystalline structure consisting of ordered, oriented chains of selenium atoms. The second crystalline form, known as monoclinic selenium, has a structure of rings containing eight selenium atoms. This form has a much higher resistivity of about 10^{12} (Ω cm) and is also photoconductive, having been one of the materials studied by Gudden and Pohl in their pioneering studies in the 1920s.[64] Monoclinic selenium crystals are extremely difficult to grow, however, and have never proved to be of any technological significance. By contrast, trigonal selenium, usually in polycrystalline form, was extensively used in the thirties and forties both as photocells and for current rectification. Rectification, a phenomenon of considerable technological importance, is the ability of a device to pass current easily for one voltage polarity but much less so for the other.[37,38] The same property is exhibited by a thermionic vacuum diode in which current between the cathode filament (the source of electrons) and anode electrode can only occur when the anode is positive with respect to the cathode. Such devices therefore provide a means of converting an ac (alternating current) into a dc (direct current). Historically, the rectification exhibited by trigonal selenium-metal contacts was one of the first examples of the use of a solid-state device to replace the more bulky, power intensive vacuum diode.[1]

The amorphous form of selenium, reported as early as 1885 by Bidwell, is a red glassy film readily made by melting selenium on a plate followed by rapid cooling, or by condensing the vapor on a cold substrate.[65] By heating to approximately 100 °C, it can be crystallized into the trigonal form; a process which can be accelerated by illumination. This effect, first observed by Bidwell, was used by Selenyi to produce photographic images on amorphous selenium plates.[66] It was also common practice in the production of early trigonal selenium crystal photocells to heat the devices to ensure the crystallization of any amorphous component, since early studies indicated that it was highly insulating with a resistivity approximately 10^{14} (Ω cm) and nonphotoconductive.[23] Contrary to these initial reports it is in fact very photosensitive, a fact that appears to have been discovered independently by several groups in the mid to late 1940s.[56,67] The only common denominator in these efforts was that each was searching for a large-area thin film photoconductor for applications in imaging devices. One of these was a group at the RCA laboratories exploring vidicon technology which involves the use of photosensors in television cameras.[39] In the vidicon picture tube or television camera, a photoconductive film on a transparent conductive substrate is charged by scanning the film with a beam from an electron gun. An optical image projected through the transparent conductive substrate induces photoconductivity in the film. The magnitude of this enhanced conductivity depends on the local intensity of the incident light. This spatially modulated conductivity which mirrors the projected image is the source of a television signal. The other effort was at Battelle where improved photoconductive layers for Carlson's invention were being sought. In fact the requirements for both technologies are very similar. The photoconductive vidicon film must have a high enough

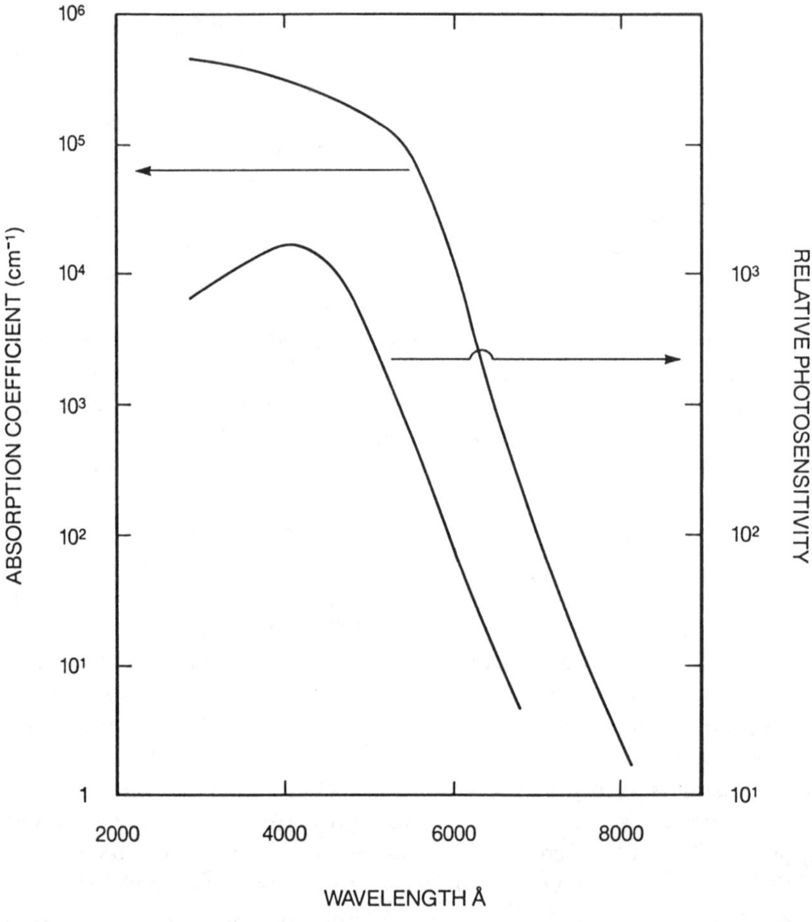

Although amorphous selenium strongly absorbs red light, there is little or no accompanying photoconductivity and the photoconductive threshold lies to higher energies than the absorption edge. As a result amorphous selenium photoreceptors are not responsive to red light. This initially inexplicable property also turns out to be intimately associated with the amorphous state.

resistivity to ensure that in the dark the voltage produced across the film by charging with the electron beam is essentially unchanged from one scan to the next. In xerography the charging is produced by ions rather than electrons and the images are not detected by scanning, but otherwise the fundamentals are essentially the same.

The combined outputs of these efforts were that amorphous selenium films produced by evaporation in a vacuum chamber were highly insulating in the dark and photogenerated electrons and holes could be created with essentially unit quantum efficiency. The spectral sensitivity extends from the ultraviolet to wavelengths of 5500 Å (green light) before decreasing sharply.[67] Somewhat surprisingly, therefore, the onset of photoconductivity occurs well into the absorption edge, so that although selenium absorbs red light very strongly it is not photosensitive to these wavelengths.[63] This was a mystery to remain unresolved for twenty more years. It was also established by both groups that free holes could move a distance at least an order of magnitude further than electrons before becoming immobilized by traps. For low light levels and high fields the photoresponse was linear in light intensity, a considerable virtue for an image transducer. With prolonged exposure to light, photocurrents were found to decrease with time. This effect, commonly observed in photoconductors and appropriately called light fatigue, is due to the accumulation of charges in traps which in turn may be associated with impurities or imperfections.[40,68]

Such properties were sufficient to allow the development of an amorphous selenium photoreceptor for the xerographic process. In particular, manufacturing processes for the deposition of the new material on drum substrates were developed, since it had been recognized in Carlson's second patent filed in 1939 that an automatic machine was best designed about a rotating drum. The basic feature of the manufacturing process involves the loading of aluminum drum or, for a subsequent Xerox product, metallized-belt substrates into a large vacuum chamber, together with amorphous selenium pellets loaded into pans called boats. After evacuation of the air with large vacuum pumps, the selenium is melted by passing a current through the boats. The selenium is thus boiled off and its vapors deposit onto the substrate surfaces. There were numerous problems to solve in terms of optimum process parameters to achieve uniform coatings of the requisite photoelectronic properties, surface quality and reproducibility. Indeed many unexpected effects were observed. One of these was that freshly deposited amorphous selenium often exhibited low charge acceptance so that for a given corotron current, the voltage a drum could be charged to was unacceptably low. Surface treatment such as physically buffing or as it was later discovered chemical treatments with chemical donors, i.e., substances with a propensity to donate electrons, could immediately provide an improvement.[69] Such unpredictable complications are often the norm in any embryonic technology and it is almost an axiom that no degree of experimentation in a laboratory can ever quite duplicate, let alone anticipate, all the possible complications that can arise from either scale-up or operation in the uncontrolled environment of the market place. Similar sorts of unanticipated difficulties were even encountered in transistor technology where a substantial understanding of its operation on a microscopic level preceded the development phase. By contrast, no understanding of the materials or processes associated with amorphous

Amorphous selenium photoreceptors are produced by the evaporation of amorphous selenium in an evacuated chamber of the type shown. In this experimental fabrication facility, a metallized belt is mounted on the mandrel located in the center of the chamber. In the lower half of the photograph, containers or boats in which selenium pellets are loaded can be seen. On heating these boats, the selenium melts and its vapor condenses to form a thin solid film on the substrate which is slowly rotated to assure uniformity. (Photograph, Courtesy Xerox Corporation)

photoconductors and, as will be discussed in the next chapter, developer materials, existed at anything approaching a fundamental level. As a result, early problems in production or performance of selenium photoreceptors had to be solved on the run using a combination of limited insight and intuition. This allowed the achievement of a sufficient degree of operational know-how to permit the technology and product development to proceed leading to the Xerox 914 copier.

Therefore, despite the successful innovation of the Xerox 914 and the critical role of the generation and transport of carriers in amorphous selenium, the first direct unambiguous measurement of these properties was not made until the late fifties. Even more striking was the fact that no coherent or sustained research by the scientific community at large had been initiated on amorphous materials. Undoubtedly a number of reasons accounted for this. First the field of solid-state physics in the forties and fifties was focused almost exclusively on crystalline solids. Not unnaturally these were viewed as being the most fruitful area of endeavor given the fact that for experimentalists and theoreticians alike they appeared the most tractable. For difficult as experiments and calculations might be, crystals did meet the requirements of periodicity so that substantial simplifying assumptions could be made and such model systems offered the prospect of both progress and fame. The dramatic discovery of transistors only served to heighten this interest and focus as more and more laboratories, particularly those in industry, competed to find another proverbial pot of technological gold in the world of crystals.[1] By contrast, the attraction of research in amorphous materials was not widely appreciated. Such materials appeared as aberrations of nature, certainly intractable, and a daunting challenge indeed to any scientist, particularly theoreticians, thinking of tackling them. In fact some reflection would raise doubts as to whether the band theory of solids, so enormously successful in differentiating between metals, insulators, and semiconductors, was even applicable to amorphous solids. For these solids lacked the very periodicity apparently required for the band theory to be valid.

It is an ironic fact that physicists spent many an hour gazing through window glass speculating if an amorphous material could have a bandgap. Window glass is indeed amorphous and yet, precisely because it is transparent to visible light, it must have a bandgap beyond the visible; depending on the glass, the absorption edge lies at about 3000 Å or 4.0 eV. If amorphous materials like crystalline solids possess bandgaps, it is natural to ask what their distinguishing feature is in terms of the band theory of solids. Impurities or lattice defects in crystals give rise to localized electron energy states lying within the bandgap. Such energy states under normal circumstances are discrete in that they have well defined energy levels. By contrast, amorphous solids possess a continuous distribution of localized

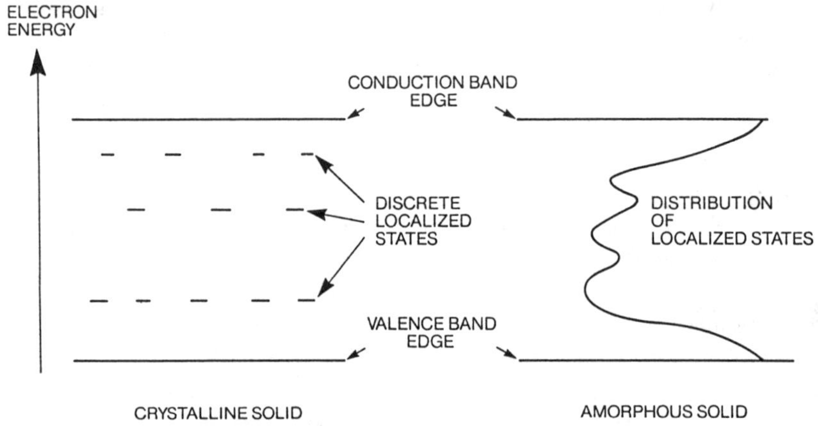

The effects of the atomic bond disorder characteristic of amorphous solids is to produce a smearing of states from the conduction bands into the forbidden gap. Smaller bond distortions, producing small energy changes, are more probable than larger ones so the density of these so-called tail states decreases rapidly with energy. They have little effect on the gross optical properties but profoundly affect electrical conduction by substantially reducing the mobility of electronic carriers. Defects such as missing atoms or impurities produce localized states lying energetically much deeper in the bandgap and typically determine the lifetime of nonequilibrium carriers such as those produced by light.

states throughout their bandgap. These localized states can be categorized into two general types. The first are associated with disorder in which the bond angles between adjacent atoms are distorted from those found in the periodic lattice of the crystal. A wide distribution of such distortions are possible ranging from small, which are more probable, to major, which are less likely so that a tailing of localized states with decreasing density from the band edges into the gap occurs.[70] In addition, impurities and defects being more severe disruptions can give rise to localized states that lie deeper within the gap. Normally these would give rise to localized states sharply defined in energy. However, they also find themselves within a random distribution of host atoms so that their discreteness in energy is lost and maxima and minima in their density result.

With this perspective, it is possible to understand the existence of absorption edges in amorphous solids since disorder preserves the notion of bandgaps while smearing the sharpness of the band edges characteristic of crystals. The apparent paradox as to why a material without a periodic structure still has a bandgap can also be rationalized by considering the extent over which periodicity is required. Detailed x-rays studies indicate that amorphous materials maintain some periodicity to about three or four nearest neighbors, which is roughly a sphere with a radius of 10 Å. Thus out to these distances from a given atom the material has, on the average, some crystalline character. Since any particular photon absorption is a highly local event in that it involves the excitation of a single atom, one can expect the absorption properties between the amorphous and crystalline forms of a given solid to be similar. Experimentally there are real and significant differences in the detailed specific absorption properties but the general position and shape can be remarkably alike. For example the absorption edge of evaporated amorphous silicon or germanium films is very much like their crystalline forms.[60,61]

The situation is totally different as far as electrical conduction properties are concerned. In semiconductors and photoconductors, electrical conduction is the consequence of the movement of electronic charge carriers, either electrons or holes, over macroscopic distances. The drift in crystalline silicon over distances of millimeters is not unknown and this represents millions of lattice sites.[30] A central question therefore is whether electrical conduction is dramatically different in amorphous materials. The most powerful tool for unraveling electronic transport in crystalline semiconductors was the Hall effect. One of the features of this phenomenon is that the sign of the Hall voltage can identify whether the dominant carriers are electrons or holes.[33] Combined with measurements of the electrical conductivity on the same sample, one can independently determine the number of carriers, their mobility and how each changes with doping or temperature. The temperature dependence of the conductivity allows the source of the thermally generated carriers to be identified. This is a great advance over measurements of the electrical conductivity alone which only determines the product of the number of carriers and their mobility. The values of mobility (expressed in cm^2/Vs) are dependent on the nature of the material and the processes that impede the motion of carriers. In crystalline semiconductors or insulators typical values range from 100–1000 cm^2/Vs (except for two compound semiconductors gallium arsenide and indium antimonide which have unusually high values in the tens of thousands). For the most widely used semiconductor, silicon, the values are approximately 1000 cm^2/Vs which would lead to a carrier velocity of approximately 10^6 cm/s or 20,000 mph in an electric field of 10^3 V/cm. However, the carrier velocity saturates close to that of the sound velocity in silicon (approximately 10^5 cm/s) as the energetic carriers lose energy to the surrounding silicon lattice.[30,39]

Contrary to intuition, the mobility of carriers in the best conductors, metals, is not exceptionally high being only approximately 30 cm^2/Vs, and the superior ability of metals to conduct electricity is due to the high density, $10^{22}/cc$, of free electrons.[28] The high conductivity of metals precludes the application of the high electric fields possible with semiconductors or insulators. As a result, under normally used voltages the drift velocity of electrons in metals is amazingly small at about 10^{-3} mph. One might well wonder how electrons traveling along a copper wire at such a speed can account for the apparent instantaneous voice transmission over a telephone. The time to respond to any disturbance in the space-charge neutrality such as when excess carriers are introduced into a solid is known as the dielectric relaxation time. This time approximately equals in seconds, 10^{-13} times the product of the resistivity of the material and its dielectric constant.[71] For copper with a resistivity of 10^{-6} (Ωcm) this time is incredibly short. Hence it is not the physical drift of any particular electron that governs the information transmission, but the disturbance of the electrical neutrality which travels as an electromagnetic wave with the speed of light in the copper wire. Such is not the case in semiconductors or insulators because of their much higher resistivities. Here the dielectric relaxation time may be longer than the time it takes the excess carriers to physically move the length of the material, the transit time; in these cases, therefore, the carriers do carry the message. This explains why the dimensions of silicon chips ultimately limit the speed of microprocessors and the current interest in higher mobility materials such as gallium arsenide.

A long dielectric relaxation time, which can be many seconds in insulators, allowed the development of a new means of studying the transport of electronic charge carriers in semiconductors and more importantly in amorphous materials. The technique, first applied by Haynes and Shockley to crystalline silicon, involves measuring the time it takes for excess carriers photogenerated by a short light flash to travel across or transit a known thickness of material with a known applied electric field.[72,73] The time-of-flight technique, as it is called, is accurately descriptive of the method. Knowing the transit time, one can determine the velocity of the carriers and thus their mobility or velocity per unit field. This mobility, known as the drift mobility, can equal the mobility determined from the Hall effect, although it is typically smaller. This is because it measures the total time the average carrier takes to make the trip. If it is held up on the way in a trap, i.e. it is immobilized for a short time, then this increases the time of the journey, thereby decreasing the average velocity and hence the mobility. By contrast, in the Hall effect the magnetic field only effects the motion of free carriers and the magnitude of the determined mobility is unaffected by traps. In the time-of-flight technique one has to differentiate between two types of traps based on the average time a carrier remains captured. This residency time is determined by the probability that a trapped

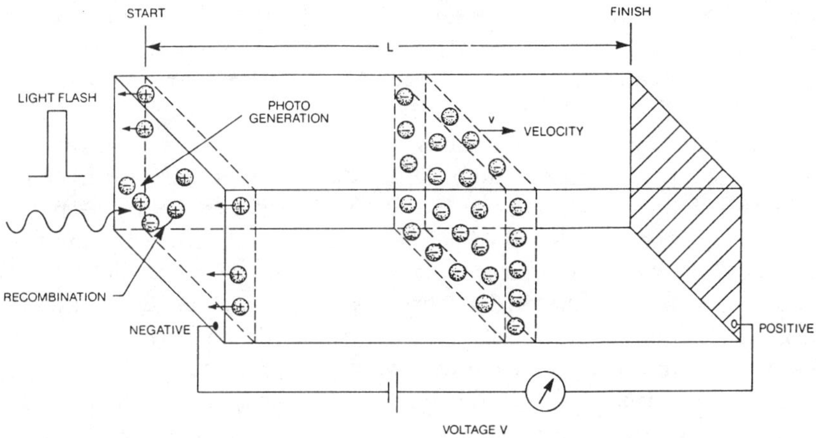

The time-of-flight method for measuring mobilities involves the measurement of the time for a sheet of charge, produced by a short flash of stronly absorbed light, to move through or transit the thickness of the sample, L. The sheet of charge drifts because of the force it feels from the applied external voltage and depending on its polarity either electron or hole transport can be independently measured. Knowing the transit time in the applied electric field, V/L, the velocity of the carriers and their mobility or velocity per unit field can be determined.

carrier can gain sufficient thermal energy to be promoted from the trap back into the conduction or valence band. This probability is related to the position in energy of the localized state within the bandgap and drops dramatically with increasing depth of the trap below the band. If the energy lies close to the band edges the trap is said to be shallow, while if the energy lies far from the band edge the state is said to be deep.[68] Deep traps are particularly significant in the operation of a xerographic photoreceptor.[74] The distinction between shallow and deep traps is best made by analogy to an automobile race. At the drop of the flag (light flash), all the cars start racing. During the course of the race the cars may stop for refueling, tire changes, etc., but they will resume. However, these stops (shallow traps) reduce the average speed. If the average number of stops and their duration are the same for each car, then it will be a close race so all the cars will pass the finish line (complete the transit) in approximately the same time. A few cars will finish somewhat behind the main body, but the spread in average speeds will be small. However, if there is a wide variation in the number of stops and their duration, then the end of the race will be strung out in time, with some cars arriving considerably later than the first to cross the finish line. To complete the analogy, some cars may blow their engines or transmissions and never reenter the race, i.e., for them the race is over since they are totally immobilized (deep traps).

In summary, therefore, shallow traps slow the carriers, reducing their average speed and effective mobility, μ; deep traps on the other hand terminate their transit. The average time before a carrier falls into a deep trap, termed its deep trapping lifetime, τ, times its velocity determines the distance a carrier moves before this occurs. The product of this mobility and lifetime is called the carrier range or $\mu\tau$ product and determines the degree of discharge achievable in a xerographic photoreceptor.[74] The reason the time-of-flight technique is so important is that the Hall effect is difficult to measure if the conductivity of a material is too low. Since the conductivity is dependent on the number of free carriers and their mobility, if either of these are low the Hall voltage is difficult to detect. If both are unusually low, as is the case for most amorphous materials, the voltage is beyond detection.

It is worthwhile to recapitulate the state of knowledge regarding the functioning of amorphous selenium photoreceptors in the mid fifties. The work of scientists at the RCA laboratories and Battelle had uncovered important facts.[56,57] First, the material exhibited high photosensitivity to light in the visible part of the spectrum, although for some reason the strong light absorption in the red part of the spectrum produced surprisingly little photoconductivity. The distances that carriers could move, particularly holes, were substantial, being of the order of tens of microns. More importantly, it had been found that operationally amorphous selenium functioned exceptionally well as a photoreceptor in the xerographic process. However, in the case of amorphous selenium there was no knowledge of the carrier mobilities or explanation of the lack of red photoresponse.

It was not until 1957 that the first direct measurement of a carrier mobility in amorphous solids was made. These measurements were carried out at the University of Leicester in England by Spear.[75] Using the time-of-flight technique, he reported measurements of the mobility of electrons in amorphous selenium. The values he found were approximately 10^{-3}cm^2/Vs; almost six orders of magnitude smaller than in crystalline silicon. As previously discussed, this drift mobility was reasonably expected to be influenced by trapping effects. Information on traps, i.e., their density and position in energy with respect to the band states can be deduced from studying how the drift mobility depends on temperature. From such studies he concluded that there were approximately 10^{20} traps per cubic centimeter lying 0.3 eV below the conduction band and that these traps were a direct consequence of the disorder intrinsic to the amorphous state. At that time, he was unable to measure the mobility of holes because

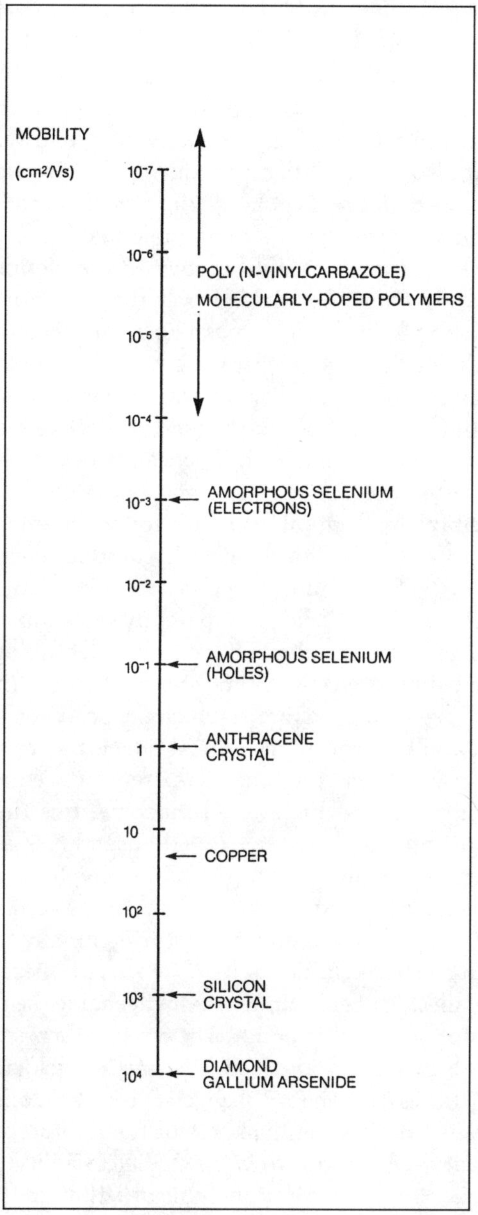

Mobilities of electronic carriers in a variety of solids. Semiconductors generally have the highest carrier mobilities, while for metals which are the best conductors of electricity electron mobilities, in copper for example, are quite modest. Amorphous solids and even molecular crystals are characterized by very low carrier mobilities.

he surmised their mobility to be much higher than for electrons and the light pulse used to photogenerate carriers was too long to allow the measurement of the faster transit times. After developing a faster carrier generation pulse, using a pulsed electron beam, he measured the hole drift mobility at 0.1cm^2/Vs which was thermally activated with an energy of 0.14 eV.[76] Spear also realized that by reducing the electric field across the selenium sample he could reduce the carrier velocity and increase the time it took the carriers to move across the sample thickness. This transit time could only decrease, however, until it equaled the lifetime of the carriers wih respect to deep traps. For fields lower than this, the carriers would move only a fraction of the sample thickness before becoming permanently trapped. This allowed a direct measurement of the deep trapping lifetime. Knowing this and the drift mobility, he was able to determine the range of holes to be about 10^{-6} cm^2/V; the value for electrons was an order of magnitude smaller. These were in complete accord with the more indirect, but correct, observations made at RCA and Battelle.

The incredibly small mobility values found in amorphous selenium compared with crystalline solids were understood to reflect the disorder of the amorphous state. Such sauntering carrier motion at first seems irreconcilable with their role in producing, apparently instantaneously, the latent electrostatic image in xerography. The resolution of this apparent paradox lies in the relationship between the time of events within the photoreceptor and the process speed, i.e., the rate at which the process is repeated per second. In the xerographic process the critical time is the interval between the light exposure in which the photocarriers are produced and the development step.[74] For a modern high speed machine, this time can be a few milliseconds (one millisecond equals one thousandth of a second). In this time, the photocarriers must transit the photoreceptor thickness of about 50 microns to discharge it and produce the charge pattern in the latent electrostatic image. Typically the photoreceptor is initially charged to about 1000 volts or an electric field of 2×10^5 V/cm. With these values, one can calculate the smallest carrier mobility that will enable the carriers to transit the photoreceptor in 1 millisecond. The voltage across the illuminated photoreceptor decreases in time so the carrier velocity progressively decreases, but this is reasonably neglected for the point under consideration. The transit time is the thickness of the photoreceptor divided by the carrier velocity, which in turn is the carrier mobility times the electric field value. This simple calculation indicates that the carrier mobility must be equal to or greater than about 10^{-5} cm^2/Vs.[74] This is a surprisingly low value, since even the low mobility of holes in amorphous selenium is four orders of magnitude larger. So carriers with a mobility of 10^{-5} cm^2/Vs, corresponding to a velocity of 0.02 mph in a field of 10^5 V/cm or almost literally a snail's pace, can suffice. As we shall see, this

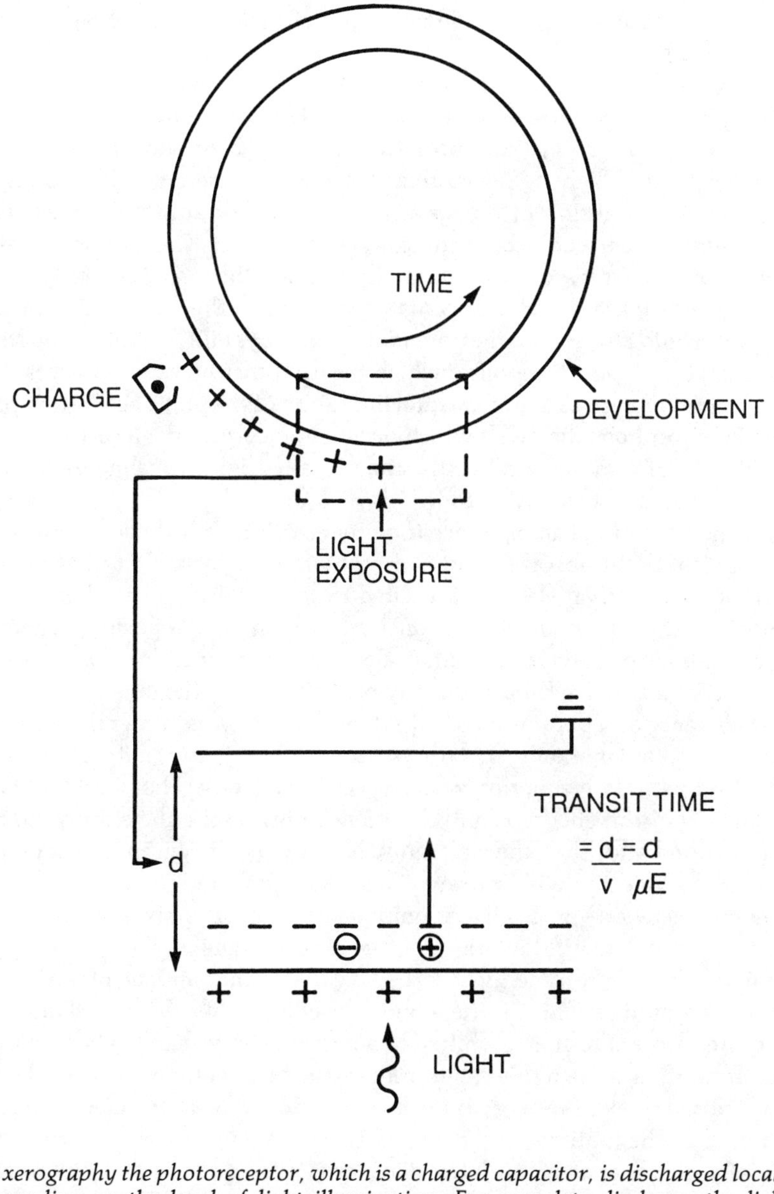

In xerography the photoreceptor, which is a charged capacitor, is discharged locally depending on the level of light illumination. For complete discharge the light-produced free carriers in the exposure must both equal the surface charge deposited in the corona-charging step and traverse the total thickness of the photoreceptor. Any deep or permanent trapping of the photogenerated charge leads to incomplete discharge, known as a residual voltage. In addition, the transit of the charge must occur in a time less than it takes for the exposed area of the photoreceptor to move from the exposure station to the development zone.

insight was to open up a whole new class of materials for photoreceptor applications.

During the course of the sixties, attention was focused on the question of the photogeneration mechanism in amorphous selenium. This was an attempt to explain the lack of photoresponse associated with the strong absorption of red light. The motivation was not merely understanding, because any extension of the spectral sensitivity of amorphous selenium that might result would be technically and product-wise of considerable significance. First it would substantially increase the sensitivity of the process by utilizing more of the available photons in the white light sources used. It would also extend the potential color copyability. It turned out that this lack of red photoresponse, which the customer literally observes, is a direct consequence of a fundamental limitation of the photogeneration process in amorphous materials which occurs in picoseconds; a picosecond is a billionth of a millisecond or the time it takes light traveling at 186,000 miles per second to travel 1 foot.

The limitation can be understood by enquiring a little deeper into the mechanism of the absorption of a photon of light. For a photon of the appropriate energy, an electron is excited from the filled valence band to an empty energy state in the conduction band. The oppositely charged electron-hole pair have a mutual electrostatic attraction, the coulomb force. This attractive force is directly proportional to the inverse square of their distance of separation. For sake of argument, we consider the hole stationary and assume only electron motion, although in reality both move and it is their relative motion with respect to each other that matters. If the absorbed photon energy is only sufficient to just excite the electron to the conduction band, then the electron will find that it is within the coulomb field of the hole and will be drawn back to mutual annihilation before it is ever completely free.[77] Such a recombination mechanism is called geminate recombination (from the latin word gemini for twins). This can only be avoided if the photon energy absorbed is equal to the bandgap plus an additional amount equal to the coulomb energy of $e^2/2\,\kappa r$. Thus the photoresponse threshold is shifted to a higher energy than the bandgap by this amount, although this can be reduced at sufficiently high applied electric fields. The excess energy is inversely proportional to the dielectric constant κ of the material and r, which is the distance that the carriers can initially diffuse apart. This latter distance is related to the mobility of carriers and is smaller the lower the mobility. Geminate recombination, therefore, is more likely in amorphous materials because of their low mobility since the lower the mobility, the smaller r and the larger the additional energy that must be provided. For photons whose energy is equal to the bandgap plus this excess energy, the photoexcited carriers are guaranteed escape and the generation efficiency is essentially unity. If the photon

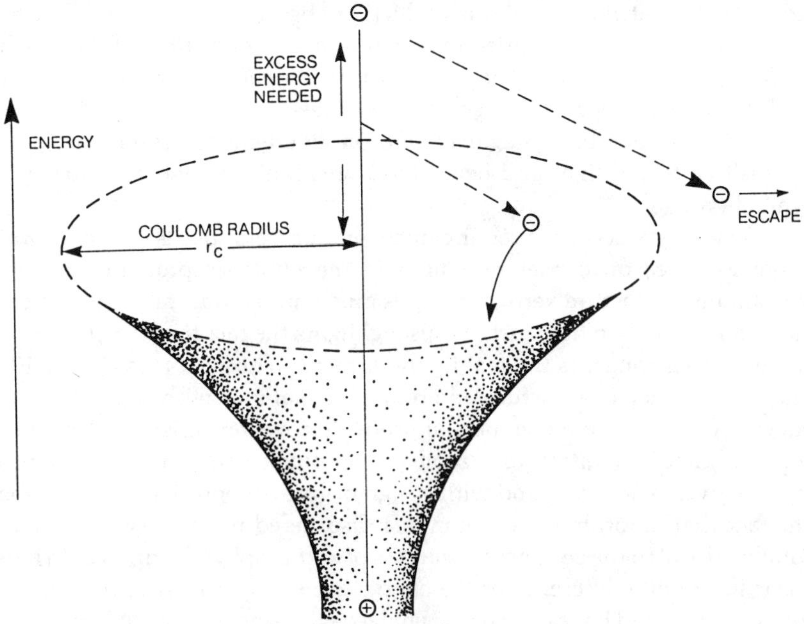

Geminate recombination can occur in low mobility solids such as amorphous materials. An electron excited by a light photon from the valence band experiences a coulombic attractive force from the positively charged hole created in the valence band. If the electron does not get beyond a distance known as the coulomb radius, r_c, it never becomes free, is drawn back into the coulomb well, recombines with the hole and no photogeneration occurs. To avoid this, the electron must acquire an initial energy from the photon which equals the band gap energy plus an excess equal to the coulomb energy; the onset of photoconductivity is therefore shifted to shorter and more energetic photon wavelengths than the absorption edge.

energy just equals the bandgap energy, the carrier pairs have a high probability of recombining before ever being freed and the photoresponse is very low. In the case of selenium the additional required coulomb energy is about 0.3 eV at low fields. This accounts for the low photosensitivity of amorphous selenium to red light even though this bandgap light is strongly absorbed. A compound of arsenic and selenium, amorphous arsenic triselenide, which has also been used commercially as a photoreceptor, does not exhibit this nonphotoconducting absorption even though it has even lower carrier mobilities than amorphous selenium.[78] This can be understood since its dielectric constant is 9 compared with 6 for selenium and the increased screening reduces the excess energy required.[79] Since it has a very comparable bandgap to amorphous selenium, arsenic triselenide is sensitive to red light and hence has a much higher photosensitivity to white light.

As we have seen, the requirements that process speeds of xerographic products place on carrier mobilities in the photoreceptor are not very demanding. In fact, in xerography it is much more important how far carriers can move than how fast.[74] This highlights the fact that the $\mu\tau$ product or carrier range, is the real figure of merit. Thus a $\mu\tau$ value of 10^{-6} cm^2/V or larger is a useful definition of a material with a good carrier range. When used in a commercial product, however, it is also absolutely essential that in addition to a good $\mu\tau$ product a material must be producible over large areas and with a high degree of reproducibility, so that the fact that amorphous selenium is a disordered material was of course vitally critical to the commercial success of xerography. Things would have turned out quite differently if the charge transport properties required by xerography could have only been achieved with single crystals. The relaxation of the periodicity therefore was a major enabling feature of amorphous materials. As we have seen, however, an inevitable consequence of disorder is the appearance of localized states within the forbidden gap. These can act as recombination centers which control the photogeneration process and shallow and deep traps which reduce the carrier drift mobility and deep trapping lifetime, respectively. Two major questions were dominant throughout the sixties. The first was related to how reproducible these materials could be made. Secondly, was whether it would be possible, from an accumulated understanding, to identify different material embodiments. It was presumed that alternatives to amorphous selenium might well offer different and advantageous properties.

The sixties and seventies saw the birth of the systematic experimental and theoretical study of the amorphous state in general, as concerted fundamental studies throughout the world were undertaken. Specifically, in the case of selenium, further insight was gained into this unusual material. Empirically it had been established that the material was remarkably reproducible. Following Spear's pioneering work, many measurements of the hole drift mobility in amorphous selenium were made throughout the world. Despite the fact that the films were made in different laboratories from different source materials, the results never differed by more than 20 percent. Such reproducibility matched that obtained in crystalline semiconductors, such as silicon, grown with scrupulous attention to perfection and purity.[80] On one level this could be explained in terms of the atoms in amorphous selenium being arranged so randomly that it would not be difficult to reproduce the same degree of randomness. As far as the hole drift mobility is concerned, this is a reasonable explanation, since the value of 0.1 cm^2/Vs is in fact determined by a trapped-controlled process. The hole alternately moves through the valence band with a free microscopic mobility of about 10 cm^2/Vs and is trapped many times in a very high density, about 10^{20} per cubic centimeter, of shallow traps 0.14 eV in energy above the valence band. Such an explanation is less readily believable with respect to the deep trapping lifetime. Extensive studies indicated that this lifetime is controlled by deep traps located essentially in the middle of the gap, i.e., about 1 eV from the valence band with a density of about 10^{14}/cc. Such a low trap density of about 1 per 10^8, the equivalent of 2 persons in the entire population of the United States, is hardly matched by the purest silicon crystal. Moreover, given this remarkably low density of defect states, one would expect significant sensitivity to the presence of impurities. Now although it is found that it is certainly possible to modify the properties of amorphous selenium by the deliberate introduction of impurities, the degree of sensitivity encountered is much less than one would predict based on this background level of localized states. The answer to the apparent anomaly lies in the nature of selenium and the specifics of its chemical bonding.[81] In fact one of the clarifying advances of research in amorphous solids in the seventies was the revelation that, despite the inherent disorder of amorphous materials, the individualistic properties of the constituent atoms are preserved.[82]

Amorphous selenium is a member of a special class of amorphous or disordered materials called glasses and is sometimes called vitreous selenium. In a glass, which is a supercooled liquid, there are sufficient degrees of freedom to allow the atoms to adjust to satisfy their bonding requirements. In other words the glass can self-anneal at room temperature to minimize the number of unsatisfied bonds or defects and related localized traps. Indeed from its earliest use, it had been empirically observed that the electrical and photoelectronic properties of amorphous selenium would change with time for several days after deposition, reflecting these thermal relaxation effects.[69]

Amorphous selenium clearly played a central role in the phenomenal success of xerography. Despite this, selenium had some definite shortcomings in terms of the future evolution of the technology. First as observed by Bidwell and Selenyi it exhibited a propensity to crystallize to the stable trigonal selenium form.[65,66] Even if this had occurred uniformly, it was disastrous because the low resistivity (10^6 Ω cm) of this form of selenium rendered it unusable as a photoreceptor. This particular problem, however, was solved by alloying the selenium with small amounts of arsenic which was found to inhibit crystallization by introducing crosslinking bonds.[69,78] Other shortcomings less easily remedied were the lack of red photosensitivity and its intrinsic lack of flexibility. From an engineering perspective, considerable virtue was ascribed to having a flexible photoreceptor and, by using a materials engineering approach, amorphous selenium belts were produced by Xerox in the seventies for use in the high speed Xerox 9200 machines. Another driving force from the business perspective, however, was the need for even cheaper photoreceptors to reduce costs and make the products based on xerography available at lower prices to the customer. The combined need for extended spectral response, flexibility and low cost were therefore major driving forces to identify alternative photoreceptor materials and devices during the sixties and seventies. The advances that came capitalized on the accumulated insights of the first generation of photoreceptors, progress in the understanding of the amorphous state in general and the relationship of this knowledge of materials and phenomena to actual device performance. Alternatives to amorphous selenium and its alloys in terms of flexible thin-films exist in organic polymers.[17] It was therefore natural that attention turned to these materials in the sixties. Polymers as a material class exist widely in the natural world, although the ability to synthesize such materials had to await the birth of organic chemistry; the first known synthesis of an organic compound from inorganic precursors was that of acetic acid by Kolbe in 1845.[10,17] Organic chemistry is so named because it deals with substances which were originally thought to derive only from living organisms; today, with some exceptions, it relates to all substances containing carbon. Polymers are a particular class of organic compounds which have distinctive characteristics. They consist of a single unit (called a monomer) connected together, much like a string of pearls in a necklace, to produce chains which may contain millions of these subunits. If these individual strands or chains tend to align parallel to themselves, either because of bonding between the chains or sometimes by being stretched, this introduces some periodicity. Such crystalline polymers, cellulose is one example, can diffract x-rays. Other polymers are totally random or amorphous in that the chains do not line up or chemically bond to each other and can be visualized as a bowl of spaghetti. Many of these amorphous polymers are glasslike in character with outstanding flexibility, transparency and impact strength. All automobile windshields use safety glass in which a thin polymer film is sandwiched between two tempered glass sheets. Polymers such as polyacrylics and polycarbonates can be substituted for ordinary glass with advantage since they are more easily molded and do not shatter when broken. All such synthetic polymers have been produced, developed and commercialized since the 1930s. Despite these desirable mechanical and chemical properties, polymers generally constitute some of the best electrical insulators known. As a result they were rapidly and widely employed for this purpose in producing electrical cables, wires and insulating films. Consequently, they were not viewed as plausible candidates for applications where electrical conduction was needed. Despite this, some rudimentary studies were initiated to study the electrical properties of organic materials in general and polymers in particular. An implicit goal was the hope that the synthetic creativity of organic chemists, together with an understanding of the organic solid-state might, by combining electronic function with desirable mechanical features, result in unique properties. As we shall see xerography is the beneficiary of those early dreams.

In the late 1940s and into the '50s, it was discovered that films obtained by dissolving aromatic organic molecules and polymers in a common solvent, which was then driven off, could be charged with a corotron and discharged with ultraviolet light.[83] By this time it had also been established that solid films of various organic dyes exhibited photoconductive effects in their visible absorption bands. The addition of small amounts of such dyes, which dissolved to form a solid solution, greatly increased the photosensitivity of these ultraviolet-sensitive films.[84] Significantly it was found that when the dyes were dissolved in the polymer matrices without the aromatic molecules, no photodischarge was observed. The first noteworthy progress in the study of electronic carrier transport in organic materials did not really occur until the late fifties. Precise measurements were made by Kepler in anthracene single crystals using the time-of-flight technique.[85] The reader will recall that anthracene films were investigated by Carlson in his early xerographic studies. Even though Kepler and LeBlanc's measurements were in single crystals, the measured mobilities of carriers photogenerated in the anthracene by a flash of ultraviolet light were only about 1 cm^2/Vs and the temperature dependence indicated no strong trapping effects. Similar results were observed for other aromatic molecular crystals. This suggested that the low mobility is intrinsically associated with the nature of aromatic molecular systems. Such crystals consist of cyclic molecules with strong covalent bonding within the aromatic ring. In inorganic materials, where the crystal building blocks are atoms, we have seen that the strong interaction between atoms when they are condensed with 10^{22} atoms/cc leads to the formation of energy bands. In molecular crystals the building blocks are neutral molecules which interact weakly, even in the solid phase, via the so-called Van der Waals bonding force; it will be seen later that this type of force also plays a critical role in the xerographic development process. This type of comparatively weak bonding, as in the case of solid neon, is manifested by the characteristic softness and low melting temperatures of aromatic organic solids. Another consequence is that the energy bands formed in such molecular solids are extraordinarily narrow and this can be shown to be consistent with low electronic carrier mobilities even in perfect, pure crystals.[86]

Similar considerations apply to polymers. For example, the absorption of light by polystyrene is little changed whether it is measured in a solid film or solvent solution. Polystyrene consists of a vinyl backbone that defines a one-dimensional chain. Attached to the chain and pointing away from it into space are aromatic rings. Such polymers, in which the monomer units hang from the chain are called pendant group polymers thus maintaining the necklace analogy. The monomer units are sometimes referred to as chromophores since they dominate the absorption and, depending on the unit, may impart color to the polymer. The absorption at the lowest absorbed photon energies is dominated by these pendant groups which exist in both films or solutions. Typically for most polymers, such absorption does not begin until the photon energy lies in the ultraviolet part of the spectrum. Many polymers do not, therefore, intrinsically absorb any visible light and are totally transparent.[87] Early on, one pendant polymer was found to exhibit unusually large intrinsic photoeffects within its ultraviolet absorption bands, without the need to dope it with a monomer.[88] This polymer called poly(N-vinylcarbazole), [PVK], has carbazole pendant groups attached to the polymer backbone and extending into space. It was found that the addition of a strong electron acceptor dopant such as trinitrofluorenone (TNF] extends the photoresponse into the visible part of the spectrum.[89] This occurs because of the formation of a charge transfer complex, PVK-TNF, which is associated with an interaction between the carbazole unit, which is a donor molecule, with the electron acceptor. As the names imply the carbazole donor molecule is willing to give a photoexcited electron to the TNF molecule which electrically is willing to accept it. This results in a new absorption band in the visible spectrum which is not possessed by either molecule alone.

These developments led to the intensive studies of PVK:TNF and PVK and in particular the transport of photogenerated charges by directly resolving their transit. When this was successfully accomplished in the late '60s, it represented a major breakthrough in the growing field of the electronic properties of polymers.[90–93] It also raised as many questions as it answered. In the first place, the measured shape of the transit pulses indicated that far from transiting the sample as a coherent sheet of charge, the carriers obviously moved with a large variety of velocities. Some arrived early, but a significant fraction arrived considerably later. The second surprise was that the velocity of the carriers increased superlinearly with field so that if the electric field were doubled the velocity would increase fourfold. The third observation was that the average carriers had an effective mobility of only about 10^{-7} cm^2/Vs even at a relatively high field of 10^5 V/cm. Fourth, only hole transport could be observed in PVK. Less surprising was

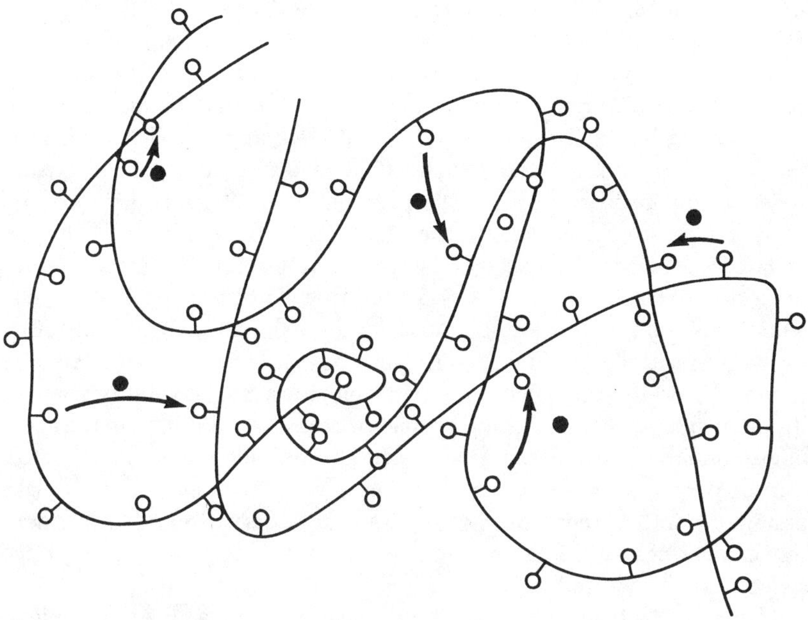

Poly(N-vinylcarbazole), PVK, is a pendant-type polymer in which carbazole groups, schematically indicated here by circles, are attached to the polymer backbone. It is also an amorphous polymer in which the constituent polymer chains are randomly entangled like spaghetti. Research established that charge transport in this polymer involves carriers randomly jumping or hopping from one carbazole unit to another either on one chain or between different chain segments. The distances between the sites involved in a hop, indicated by the arrows, are variable because of the random structure of amorphous polymers; a wide distribution or dispersion of carrier velocities is therefore expected and observed.

the fact that from the temperature dependence the mobility and therefore the mechanism of motion involved some activation energy.

The mobility observed is approximately a billion times smaller than in crystalline silicon. More graphically, a hole moving in PVK under a field of 10^5 V/cm takes about 5 minutes to move an inch. However, as discussed earlier, the speed of carrier motion is only limiting in xerography if the transit time of the carriers across the photoreceptor is shorter than the interval between exposure and development. The more critical question is how far carriers can move in a polymer such as PVK. The answer is tens of microns. Transit times as long as 20 seconds have been measured using the equivalent of a kitchen timer and constitute some record for the observation and detection of electronic carrier motion through a solid. Thus the deep trapping lifetime in this material is at least minutes! Scientists were thus faced with understanding such extraordinary behavior of electronic carriers in this pendant polymer either in terms of the general framework of existing knowledge or perhaps the postulation of new concepts. One immediate issue to be faced was whether the assumption that this represented the motion of electronic carriers was even true. The polymer material was certainly not of electronic grade purity and the films were cast from solution with inevitable residual solvent, certainly at levels of many parts per million. Given the extremely low drift velocities, the possibility that the motion of charge observed could be due to ionic drift, as occurs in the alkali halides, had to be considered. A characteristic of ionic motion is the occurrence of polarization and decomposition effects at the electrodes. By the passing of photocurrents for days without the observation of such effects, these explanations were discounted and the observed transport interpreted as electronic in nature.[93]

Thus an electronic transport mechanism which was field dependent, thermally activated and proceeded with extraordinarily long transit times and lifetimes had to be explained, together with the occurrence of a wide distribution of transit times. Since PVK was quite exceptional among polymers in its manifestation of observable electronic charge transport, it was not unreasonably assumed that its structure played an important role. Since by definition all polymers have backbones and many not distinctly different from PVK, attention focused on the role of the carbazole pendant groups. There was a general consensus that the transport probably involved the hopping of the carriers from one carbazole unit to the other. More contentious and less clear was whether the polymer backbone played any active role. Two scenarios thus existed. In the first, the carrier moved along the backbone with a respectably high mobility, became localized at a carbazole unit at a chain end, where it became stuck. After a considerable period of time it hopped with thermal assistance to a nearby carbazole unit on a different chain and so on. In the other view, the motion only involved hopping between the carbazole units and no transport occurred along the vinyl backbone whose only function was to provide a scaffolding to which the carbazole units were fixed.

Now to a solid state scientist trying to understand the extraordinary experimental results of charge transport in polymers, this was an important scientific question in its own right. However depending on the answer, the technological implications for xerography were enormous. For if the vinyl backbone were not involved, all that was important was a random ensemble of carbazole rings. In this situation, one could molecularly engineer a layer with the necessary charge transport properties by doping inert polymer matrices, chosen for desirable mechanical properties, with appropriate aromatic molecules at a sufficiently high concentration.

To test these concepts a molecule essentially identical to the pendant group in PVK, N-isopropylcarbazole, was therefore dissolved with an otherwise electrically inert polycarbonate polymer using a common solvent.[94] Polycarbonate polymers have amazing impact strength and are commonly used as a substitute for glass in places such as bus shelters to thwart would-be vandals. Having prepared such films, the time-of-flight of carriers photogenerated by light in the absorption bands of the dopant carbazole molecules was investigated. The results showed a field dependent, thermally activated transport with a wide distribution of velocities and extraordinarily long transit times. In other words, in terms of charge transport the N-isopropylcarbazole-doped polycarbonate polymer was essentially indistinguishable from PVK. A major and significant difference did exist that was both scientifically and technologically enabling. The efficacy of charge transport depended, and depended strongly, on the concentration of the added molecule. It was assumed that the N-isopropylcarbazole and polymer matrix formed a molecular solid solution in that it consisted of a distribution of isolated molecules. This was confirmed even for dopant concentrations of approximately 10^{20}/cc by high power microscopy and x-rays analysis; when the solubility limit is exceeded, phase separation occurs and the films become visibly cloudy. For a solid solution it is easy to calculate the average separation between the dopant carbazole molecules and relate the carrier velocity for a given field with the separation between molecules. It was found that the carrier velocity increased exponentially as this distance was decreased by increasing the dopant concentration. At the highest concentrations, which correspond to average separations between the carbazole units of 10 Å, the transit times of photogenerated carriers were essentially those measured in PVK. Very similar results were obtained in related studies of the PVK-TNF charge transfer complex.[90,95]

Thus it was confirmed that the transport in this system and, by analogy, in PVK involved only the carbazole rings. As we have seen, in organic crystals composed of closed aromatic rings arranged on a periodic lattice the interaction between the molecular building blocks is so weak that even in this optimum configuration the band theory of solids is marginally relevant at best. In a random molecular system such as a molecularly-doped inert polymer it has no relevance at all. The electrical properties are controlled by the dopant molecules which act as purely self-contained localization worlds. Such a neutral molecule acquiring an excess carrier becomes a molecular ion. There will be some probability that this charge, at a given temperature and applied field, will hop and move spatially to

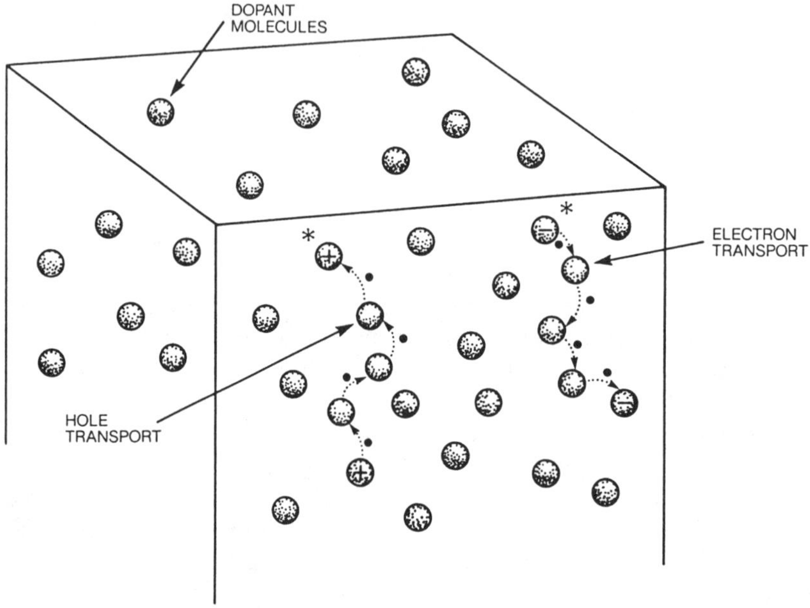

*Polymer systems can be molecularly engineered to transport charge by doping a polymer which itself does not conduct charge with electrically active molecules. In such systems, consisting of a random distribution of such dopant molecules frozen, as an example, in a polycarbonate matrix, charge transport occurs by hopping from one dopant molecule to another. The achievement of efficient transport requires that the dopant molecules be randomly dispersed with molecular separations of less than about 10 Å; this requires solid solutions of 10^{20} dopant molecules per cubic centimeter in the host polymer matrix. If the neutral dopant molecule is donor-like, that is its chemical nature is to donate an electron, then hole transport is acquired; if the dopant molecule is acceptor-like the system transports electrons. (To trace the transport steps, start at *.) Such systems, derived from an understanding of charge transport in amorphous solids in general and polymers in particular, are now in wide use as commercial photoreceptors.*

a nearby neutral molecule depending on their mutual separation. In an interesting melange of concepts, we have in these systems therefore electronic conductivity involving ions.[94,95] In the absence of any light shining on the system, a random array of unexcited neutral carbazole molecules exist. On exposure to light, an electron is photoexcited from the molecule and, following Faraday's nomenclature, a molecular cation is produced.[14] An electron can hop from a neighboring neutral molecule to this cation so that the original cation becomes neutral and the originally neutral molecule becomes as cation and so on. In effect the hole on the molecular cation jumps to a nearby neutral molecule and in one respect it appears that the cation is moving except that there is no mass motion and the phenomenon is controlled purely by electronic transitions. However in an interesting link to chemistry, the loss of an electron by a molecule is called an oxidation process while the gain of an electron is called a reduction process and thus electronic transport in a disordered molecular system can be described as a reversible solid state chemical redox reaction.[94] It is interesting to speculate on what Faraday's reaction to these developments would have been.

This framework of conceptual understanding was further strengthened by its ability to explain at least qualitatively most of the other features observed in the charge transport. Thus the wide distribution of transit times can be accounted for by the fact that the likelihood of a hop from one molecule site to another is exponentially and therefore sensitively dependent on the intermolecular separation. Since the system is disordered there will be a distribution of such distances, and hence a distribution of effective velocities is not only explained but expected.[96] The need for energy to be supplied for a carrier to hop, evidenced by the fact that it is a thermally activated process, can be visualized by the following analogy. When an excess carrier lands on a molecule, it can interact with the constituent atoms of the rather large, floppy molecule and produce some distortions, as if one dropped a small heavy object onto a soft pillow. The molecule adjusts by polarizing and the electron sinks into an energy well and requires activation to reescape. The entity of the captured electronic carrier and the resultant polarization is termed a polaron.[97]

It will be recalled that in the carbazole-polycarbonate system, as in PVK itself, only hole transport was, and ever has been, observed. One explanation could be that it is some specific property of the carbazole unit itself or some limitation in observation. There is however a more fundamental explanation. The neutral carbazole molecule is donor-like in character in other words it prefers to donate rather than accept an electron and therefore no mechanism exists for an electron to move through an assembly of neutral donor molecules. Indeed a large number of donor molecules doped into polymer matrices in this way never yield electron but only hole transport. A predictable consequence of this picture is that if a neutral, acceptor-like molecule, one which prefers to accept rather than donate electrons such as trinitrofluorenone, were used as the dopant it would be expected to only yield electron and not hole transport. Such experiments have been carried out and fully confirmed this prediction.[98,99]

Another unusual feature of charge transport in these systems is the extraordinarily long lifetimes with respect to deep traps. This comes about in two ways. First, a hole hopping between donor molecules only encounters a deep trapping site if there are sufficient impurity molecules which, having a substantially lower ionization potential, are much more donor-like. It turns out that not many molecules fit this description. Studies have been

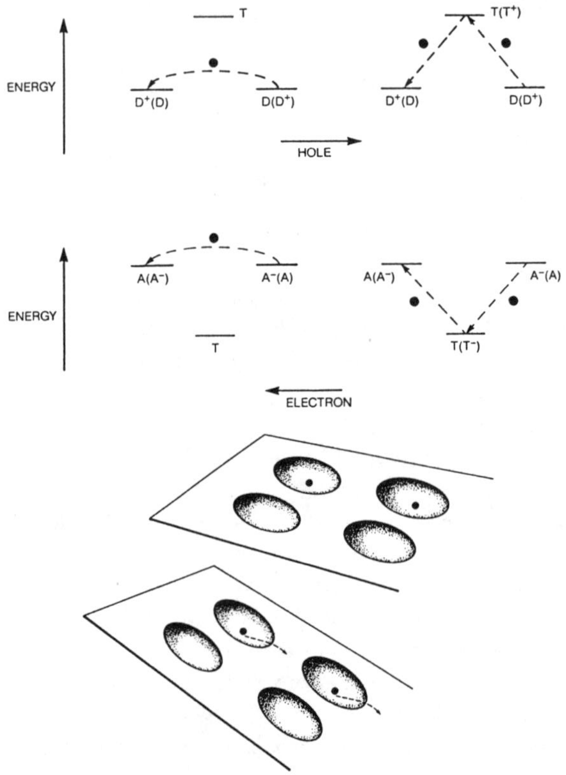

Top: *Hole transport in amorphous molecular systems involves the transfer of an electron from a neutral donor molecule to a nearby positively charged molecular cation which then becomes neutral and the neutral molecule becomes a molecular cation and so on. Ions are intimately involved in this transport process but, since there is no motion of any* specific *ion, this is not ionic conduction but truly electronic. The most efficient charge transfer occurs when the states involved have the same energy. If relatively low concentrations of impurity molecules are present which, in the case of hole transport, have more electron-donating character they can compete with the dominant transport dopant molecule and act as traps as illustrated.* Bottom: *In amorphous molecular systems each transport site is a molecule which, built from atoms, have considerable internal structure. An excess charge hopping onto such a molecule can interact electrostatically and distort it; this so-called self-trapping is equivalent to producing a potential well in which the carrier sinks. Further motion of the carrier requires that an energy equivalent to the depth of this well be provided although this can be reduced by the application of an electric field. A useful analogy, illustrated here, is to think of the carrier as a small ball rolling across a muffin-tin held horizontally. If the ball falls into one of the depressions it further motion can be aided by tilting the plane of the tin; an applied voltage serves to tilt the potential distribution in an analogous fashion. Such thermally-activated transport also means that carrier velocities increase rapidly as the temperature of the solid is raised.*

made in which a system like a carbazole doped polymer has been simultaneously doped with slightly lower ionization potential molecules.[100] The effect is to further reduce the average hole velocity, i.e., such molecules act as shallow traps. In any case, the moving carrier is only exposed to traps as it hops from one site to another, since it is safe while resident on a site. The residency time on a site, which reduces its average velocity, also decreases the probability that it will be deep trapped.

Finally is the question of the superlinear dependence of the carrier velocities on field. Although the detailed explanation of this effect is still unresolved, a qualitative explanation is possible. The carrier can be visualized as a small ball sitting in a depression, representing the polaron well, surrounded by adjacent molecular sites. The ball will normally remain where it is unless it can acquire sufficient energy to move out of the depression to another molecular site, where another polaron well is formed. However an applied electric field reduces the effective height of an energy well by tilting it and thus strongly increases the probability of the carrier's escape.[101] A similar phenomena explains the field-dependent photogeneration in amorphous selenium where the electric field effectively reduces the coulomb binding energy of the geminate pair. For all these phenomena the effective activation energies involved becomes smaller at higher fields.[77]

The molecular engineering of low-cost polymeric systems to make otherwise electrically inactive polymers, with desirable mechanical properties, conduct charge was certainly of the highest scientific interest. These systems opened up the possibility, which has been fully exploited, to explore the general question of intermolecular charge transfer and its relationship to molecular properties and structure; questions which are not insignificant to chemistry, biological systems and other technologies such as electrochemistry and photochemical solar systems.[87] Moreover they have served as important model systems in the study of the dynamics of charge transport mechanisms, such as hopping, that are generic to disordered systems whether inorganic or molecular. Their significance lies in the fact that the localized states are molecules and the type and density of localized states can be chosen by selecting the right bottle, as it were, and weighing out the appropriate weight of organic materials with which to dope.[95,98,100]

From the perspective of xerographic technology, however, such molecularly-engineered polymers were by themselves of limited value. Most of the active dopant molecules absorb only in the ultraviolet part of the light spectrum which makes them of little value as a photoconductor for use as a practical photoreceptor. This was clearly recognized at the outset and in parallel with the developments just described, efforts were expended on devising ways to make these systems responsive to visible light. The first was to incorporate into the doped matrix a visible light absorbing dye. This can be done in two ways. The dye molecules can be dispersed on a molecular level at a large enough doping concentration to ensure sufficient absorption of visible light. In this approach, one has a situation similar to a molecularly-doped polymer in that a photoexcited dye molecule, in order to sensitize the system, must transfer charge created by visible light to a neighboring neutral transport molecule. Several problems were encountered with this approach. First the overall photogeneration efficiency was low, since this was controlled by the probability of such transfer and this is limited by the same parameters as the subsequent transport itself, viz, field, activation energy and separation between the dye and a neutral transport molecule. If these limitations were not enough, there are relaxation processes that occur within the dye molecule that compete with the photogeneration or freeing of a carrier. Furthermore it was found that unexcited dye molecules could act as traps, thus impeding the transport process itself.[102] The alternative was to use an insoluble dye, referred to as a pigment dispersed as macroscopic particles throughout the matrix doped with transport molecules. Since the photogeneration step occurs within the microcrystalline dye particles and is not controlled by a dye-transport molecule distance, such systems can exhibit quite high photosensitivity. Of course, at the surface of such a particle, the transfer of the free photogenerated carrier is controlled by this distance. A major shortcoming of this approach is that the light absorption and photogeneration occurs throughout the layer. Since the transport medium only transports one sign of carrier (either electrons or holes), this results in one sign of carrier produced throughout the layer with no way of being removed. This is an unacceptable perturbation on the discharge characteristics and the ability to reproduce them on charge-expose cycling of the photoreceptor.

A different option is to physically separate the two functions involved in the discharge of a xerographic photoreceptor by using two distinct layers.[91,93,103] One layer serves as a visible absorbing source of photogenerated carriers. This layer is in physical or contiguous contact with the second whose sole function is to transport photogenerated carriers which have to be injected into it. The photogenerator layer thickness, typically < 2 microns, need only be such as to provide sufficient light absorption. The transport layer thickness, approximately 20 microns, is chosen to ensure that the total charging voltage of the device is compatible with that required by the xerographic development system to be used. In order to protect the relatively thin generator layer, it is generally put down first and overcoated by the transport layer which is transparent to visible light.[103]

The dual-layer polymeric photoreceptor serves as a good example to illustrate the all-important relationship and relative roles of research,

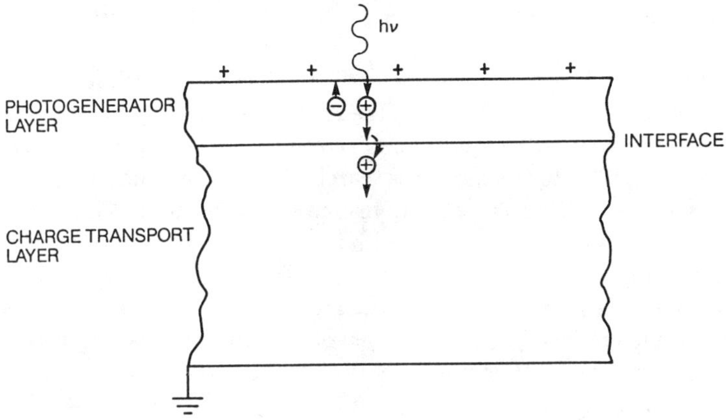

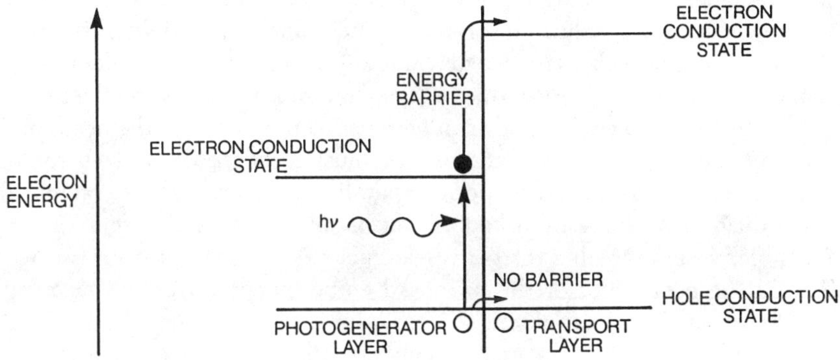

In dual-layer photoreceptors, photogeneration and charge transport functions occur in separate layers. As a consequence a new phenomenon, charge injection, must be considered, since charge photogenerated in one layer must now be injected into the transport layer. This must occur with such efficiency that essentially all the carriers arriving at the interface are injected. This also requires an alignment of energy levels for the appropriate carrier in the two solids. It is remarkable that such systems exist in which quantum particles cross interfaces that are far from atomically clean with no loss. In practice the generator layer is deposited first and then overcoated by the transport layer which also functions as a protective coating.

product planning, engineering and marketing in the innovation process. Each plays an indispensable role but has different imperatives and perspectives which must be resolved and melded to achieve a successful innovation.[9] In the hierarchy of innovation, the functions are listed in increasing order of the potential for independent action in the innovative process. Thus it is essentially impossible for research alone to produce an innovation; an invention, discovery or idea, yes; an innovation, no. At the other extreme, it is possible for marketing with an existing portfolio of products to devise new marketing strategies, pricing, leasing versus sales, etc., that can profoundly affect the competitive scenario. It is also possible for product planning and engineering to solve a problem, or initiate a design change that can result in a product enhancement by doing new things in an advantageous way; all problems and their solutions not necessarily being limited by a lack of technical insight.[2,9] In an industry which is technology based and highly competitive, however, leadership can depend critically on research progress. A new material, process or device created or discovered in a research laboratory will, however, only succeed if it is perceived to fill a market need. This may be one previously unsatisfied, or an improved way over existing solutions or more uncommonly, and by definition difficult to predict, able create a new market. Here the product planners or market experts play a critical role in understanding and assessing the potential of the research effort. They can help define more clearly the needs and requirements that the research concept must meet in areas of customer needs, such as product function, capability and price. Next technical feasibility must be established in terms of the extant limits of the technology, and whether further research can refine or improve it further. Finally the economics of the proposed embodiment in terms of performance, reliability, manufacturing scale-up and cost must be assessed. Finally all of these factors must be concurrently considered in the context of a competitive environment.

There were perceived market needs that could in principle be met by dual-layer organic photoreceptor technology. In the first place was the issue of cost. Organic materials are comparatively cheap and using solutions lend themselves to the continuous coating of rolls or webs of photoreceptor. This is inherently more cost effective than the batch process involved in the vacuum deposition of amorphous selenium on drum substrates. Such a polymeric-based photoreceptor would be flexible and therefore producible in the form of belts. For high speed copying of documents, where slit scanning involving mechanical motion of exposure lamps becomes more difficult, belts allow the imaging of a full document by a single flash exposure. Like a flash camera, this freezes a moving document. In addition, flexibility permits improved paper handling at high speed by allowing paper stripping at a smaller diameter roll. Finally, it allows the principle of tailoring the photoreceptor to a particular application. For example, with the emergence of the laser, discussed more fully later, the photogenerator part of the device could be changed depending on the laser spectral output. Such a photoreceptor could enable new innovative products with enhanced and new features at lower costs in an increasingly competitive industry. By the late sixties and early seventies other companies, including both U.S. and Japanese, were entering the market with xerographic products. All of these companies were cognizant of the market and its needs and had significant research efforts of their own in the field of xerography and the growing potential of organic photoreceptors. There was no question that from a purely scientific perspective the problems posed by layered polymeric-based photoreceptors were intensely interesting; but these could not be considered or pursued out of context. The product advantages such devices offered have been outlined above, but neither product planners or engineers are willing, ab initio, to yield on any relaxation of performance over existing technologies. Not unreasonably, they argue that any new technology must demonstrate a significant advance in all aspects of performance or cost, or at least exceed in some and match the rest. Rarely, unless one advantage is enabling, will any shortfall in another be acceptable. Moreover, the costs of one change rippling through engineering design and manufacturing may, worse still, induce a snowball effect by requiring additional changes in other subsystems. With this backdrop, the issue was whether carriers could be photogenerated in one layer and injected into another with essentially no loss so as to match the unit photogeneration efficiency already achievable with amorphous selenium. Furthermore this injection would have to occur at high speed for hundreds of thousands of cycles.

The initial prognosis contained both good and bad elements. The early work at RCA in the fifties on photoconductivity in amorphous selenium for its potential applications in vidicon technology described the use of a thin layer of antimony trisulfide deposited on top of amorphous selenium.[104] Unlike amorphous selenium, antimony trisulfide exhibits photoresponse to red light. Experiments showed conclusively that holes generated by red light in the sulfide layer could be injected into and transported across the thicker selenium transport layer to yield a hybrid structure with red light sensitivity. Unfortunately no quantum efficiency measurements were mentioned, so no conclusion could be drawn as to the efficiency of the injection process. It was also clear from earlier work that carriers photogenerated in dye molecules could be injected, at least to some degree, into aromatic molecules dispersed in polymer matrices. Again the lack of quantitative measurements did not allow the assessment of the efficiency with which this took place. From a conceptual perspective, moreover, it is possible to construct a rather pessimistic scenario concerning the possibility of achieving injection of electronic carriers without significant loss. The requirement is for holes, as one example, photogenerated in a generator layer to find themselves isoenergetic with transport states in a transport layer, since with a mismatch in energy between the two transport states only a fraction of the carriers can be injected. In addition to this constraint, there is the nature of the interface between the two materials. Thus polymers are almost impossible to produce in electronic grade purity and in any case the films coated from solution certainly retain some residual air and solvents. To complicate matters further, the interface produced by vacuum or solvent coating a generator layer would, by semiconductor technology standards, be more macroscopic than atomic. Given these concerns the likelihood of a quantum particle such as an electron or hole passing through such an interface without substantial scattering seemed remote. Now despite these seemingly well-based reasons why efficient injection would not be expected to occur, it is now known that in fact it does. This is not to say that these were not legitimate concerns, as indeed they were, but rational reasons as to why something may not work are an insufficient basis for concluding it will not. As in all research the ultimate test is to do the experiments and more importantly interpret them correctly. The latter important caveat is made because the initial experiments to quantitatively measure the injection efficiency in dual layer photoreceptor structures appeared to confirm the worse predictions. To study this injection phenomenon, a system was chosen in which there was some understanding of the individual constituent parts. Thus amorphous selenium was chosen as the generator layer and PVK as the transport layer.

Studies of this system immediately revealed disturbing observations. It was found that although the efficiency of injection could approach unity, i.e., every carrier photogenerated in the amorphous selenium could be injected into and transported through the PVK, this was only true at high fields. As field decreased, as it does in the xerographic process due to the discharge of the photoreceptor, the efficiency of injection appeared to decrease at an alarming rate.[93] This decrease was in fact proportional to the fourth power of the field across the photoreceptor. Thus after the field had decreased by a factor of three the injection efficiency had dropped by a factor of 80 so that only about 1 percent of the photogenerated carriers could be injected. For an operational photoreceptor this was completely unacceptable. The challenge was to understand and, if possible, rectify this problem. The most reasonable explanation involved an energy difference between the transport states in the amorphous selenium and PVK leading to an energy barrier to injection. Such a barrier could be field dependent so that it was sufficiently lowered at high fields to be unimportant.

A detailed analysis of this possibility, together with additional diagnostic experiments, suggested that as reasonable as this might be, this

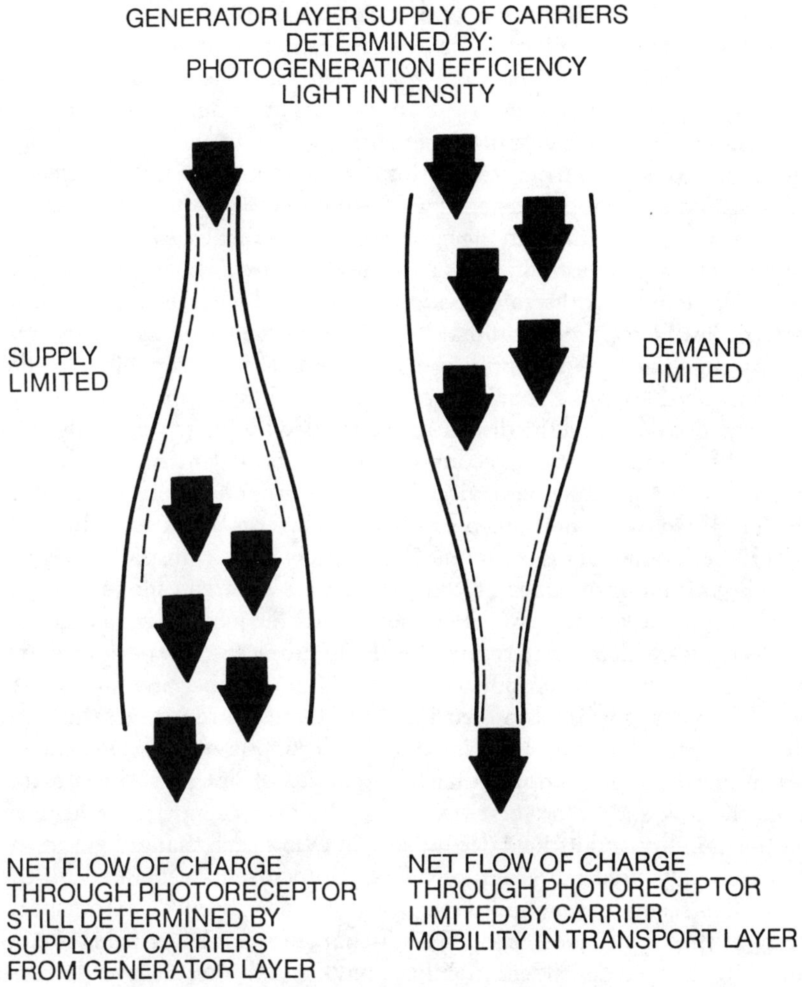

Even if there is perfect alignment of energy levels, however, another injection limitation may come into play. The supply of carriers from the photogenerator layer, determined by the photogenerating light intensity, should not exceed the ability to move the charge from the interface. This is determined by the mobility of the carriers in the transport layer. If the mobility is less than about 10^{-6} cm^2/Vs, *then this limitation comes into play at light intensities normally encountered in the xerographic process; in these circumstances, the discharge rate no longer depends on the light intensity and the photoreceptor ceases to be an image transducer.*

could not be the explanation and the cause must lie elsewhere. The understanding of the true cause can be understood by analogy to what in economics is called the law of supply and demand. The device consists of a generator in which photogeneration of carriers determines the supply rate of carriers while the transport layer's function is to move this supply. Assuming no energy barrier to injection, two extremes can occur. The first is that the rate of supply (determined by the photogeneration rate in the selenium and proportional to the light intensity) is always less than the rate of removal by transport. In this case, the flow of charge in the device is solely determined by the rate of supply and the discharge rate is said to obey the law of reciprocity in that it is proportional to the light intensity. If, however, the rate of supply of carriers exceeds the capability of the transport layer to move charge, then a bottleneck develops. Charge piles up at the interface, and the discharge, controlled by the transport properties of the transport layer, becomes sublinear in light intensity which is unacceptable in an imaging device. The next question is what determines which of these two conditions prevail. The answer is, for a given light intensity, the determinant is the mobility of carriers in the transport layer.[74,93]

The maximum number of charged carriers than can move across a photoreceptor in a single transit cannot exceed the number of charges on the electrodes which is determined by the photoreceptor capacitance and its surface voltage. The product of this maximum number and their velocity, which is proportional to their mobility, therefore represents the maximum current that can be passed.[39,40,71] From previous independent measurements, it was known that the mobility of holes in PVK is extraordinarily low and even worse decreases with decreasing electric field. In fact the mobility and its field dependence in PVK is such that at light intensities normally used in xerography, the photocurrents generated by the amorphous selenium generator layer are larger than the PVK can transport, except at the very highest fields. The discharge rate at lower fields is controlled by the field dependent mobility in PVK and this results in a decreasing injection efficiency and mimics the existence of an energy barrier to injection. There is indeed an injection barrier but it arises from a mismatch in the ability of charge to flow at the same rate in two contiguous solids and not because of an energy level mismatch.[93] For this reason, PVK is not used in dual-layer photoreceptors, although it played a critical role in the understanding of charge transport in pendant group polymers and dual-layer device operation. It should be added that barriers due to energy level mismatch can occur depending on the materials and have been extensively studied. The fact remains though that the identification of the transport limitation as a source of injection limitation was a major step forward. These kinds of considerations have therefore to be taken into account in the choice of the molecule used in the fabrication of molecularly-doped

An example of a dual-layer organic photoreceptor consisting of a thin photosensitizing layer overcoated with a much thicker molecularly-doped charge transport layer. The low cost of polymeric materials has led to an increasing use of organic photoreceptors and combined with their intrinsic mechanical flexibility allows the production of belt photoreceptors of the type shown here. The required electronic functionality of the molecularly-doped polymer depended critically on insights gained from research into the mechanisms of electronic transport in disordered materials. (Photograph, Courtesy Xerox Corporation)

transport layers. Such layers are now widely used by copier companies around the world and constitute a remarkable industrial use of molecular engineering and the first widespread commercial exploitation of the electronic properties of polymers.

In addition to their use in transport layers, organic materials are now increasingly used as the photosensitive materials in the photogenerator layer. Again, cost and the ability to capitalize on the synthetic skills of organic chemists to produce an array of materials with extended spectral photosensitivity are driving forces.[105] By using these materials in pigmented form, it is also possible to have a unified manufacturing process using film coating technology. It has also proved a more fruitful avenue to achieve infrared sensitivity with materials such as vanadyl phthalocyanine.[106] Sensitivity to infrared light is critical to the utilization of solid-state semiconductor lasers, a subject that will be enlarged upon later.

The most recent development in photoreceptor science and technology has been the investigation and utilization of hydrogenated amorphous silicon. This is a particularly fascinating story because of its connections to the phenomenon of gaseous discharges initially studied by Faraday and which played such a pivotal role in the understanding of atomic structure. Being an amorphous form of crystalline silicon which is so dominant in semiconductor technology also gives it some additional piquancy. Amorphous silicon can be produced analogously to amorphous selenium by the condensation of silicon vapor produced by evaporating silicon in a vacuum system. The major difference is that the higher melting point of silicon means that much more energy is required to do so. Amorphous silicon produced in this way turns out to be a semiconductor with a resistivity of about 10^3 (Ω cm) and is not particularly interesting either scientifically or technologically. In fact it has never received much attention.

In 1965 Sterling and Swan at the Standard Telecommuncations Laboratory in England almost by chance discovered another form.[107] They were trying to grow epitaxial single crystals of silicon by using a gaseous discharge or glow discharge of silane gas (SiH_4). They found that a brown, reddish film deposited on the quartz tubes being used. On examination, they found that the film was amorphous silicon containing a considerable amount of hydrogen and that it was surprisingly resistive (about 10^{12} Ω cm). Moreover, they determined that it possessed considerable photosensitivity. It was not until about five years later, however, that intensive research on this material was undertaken.[108] During the early 1970s, due mainly to Spear and his co-workers at the University of Dundee, extensive systematic research led to the recognition that this material was significantly different than the other known amorphous materials.[109] It was established that the resistivity and photosensitivity were a function of the substrate temperature employed and this was related to changes in the density of localized states, due to unsatisfied or dangling silicon bonds, in the gap. Investigations showed that the density of these states was modified by hydrogen, derived from the silane gas by its dissociation in the glow discharge, becoming attached to these dangling bonds. Since the material contains several atomic percent of hydrogen, it should be viewed as a silicon-hydrogen alloy. Under conditions of low discharge power and a substrate temperature of 230°C, this density of states associated with dangling bonds could be minimized. This immediately suggested that under these conditions the hydrogenated amorphous silicon might then be doped in analogy to crystalline silicon. By admixing gases such as phosphine (PH_3) and diborane (B_2H_6), it was established that this was indeed the case and that phosphine doping produced n-type amorphous silicon and diborane produced p-type material. Thus for the first time it was possible to produce an extrinsic amorphous semiconductor.

the production of p-n diodes, field-effect transistors and similar solid state electronic devices. Unlike crystalline silicon which is commercially limited in size to wafers of about 5 inches in diameter, the amorphous devices could be produced over essentially limitless dimensions.[110]

This obviously sparked considerable interest in the potential technological applications of this new extrinsic amorphous semiconductor. The early 1970s were a period when the oil crisis sharpened everyone's awareness of the ultimate decline of fossil fuels. This precipitated substantial investment in research for alternative energy sources with solar energy being a prime candidate. In particular the technical feasibility of solar photovoltaics, the production of electricity directly from sunlight, had been demonstrated in the early 60s in crystalline silicon but the major drawback was cost. Hydrogenated amorphous silicon, seen from this perspective, was an interesting and potentially attractive option.[111,112] The ability to produce photovoltaic cells over essentially limitless dimensions would involve substantial cost savings. The absorption properties of this material, moreover, were significantly different from crystalline silicon, principally because of the incorporation of hydrogen. As a consequence this form of amorphous silicon effectively absorbs all the visible solar light in a thickness of about 1 micron, whereas in the crystalline case the corresponding thickness required is about 300 microns. Apart from the materials aspects, there were significant differences and advantages in the process by which amorphous silicon was made. The precursor materials were gases and the gas pressures, flow and mixtures were modified by valves which could lead to easy automation by computer control. Moreover it was a low temperature process at about 230°C versus 1400°C used in silicon crystal furnaces. With all of these advantages an explosion in the technological exploitation of this material occurred in the late '70s to early '80s. Although the apparent diminution of the energy crisis has lessened the urgency of solar energy research, this initial work generated other spin-offs. One example is the rapid growth of the use of amorphous silicon solar cells in consumer electronics, i.e., powering watches, calculators and the like.[113]

Given the requirements for solar cell applications, i.e., high photosensitivity to visible light, good transport properties and producibility in large areas at low cost, coupled with the high dark resistivity, the reader will understand that these are precisely the criteria for a xerographic photoreceptor. Thus in the latter end of the 1970s research was initiated on this application.[114,115] Although the use of amorphous silicon in solar cells implied their suitability for photoreceptor applications, there are distinct differences in the specific requirements for the two technologies. Although for solar cell applications the film should be 1 micron or less, because of the requirements of the development subsystem photoreceptors need to be much thicker (about 20 microns). A solar cell is an integrating device in which

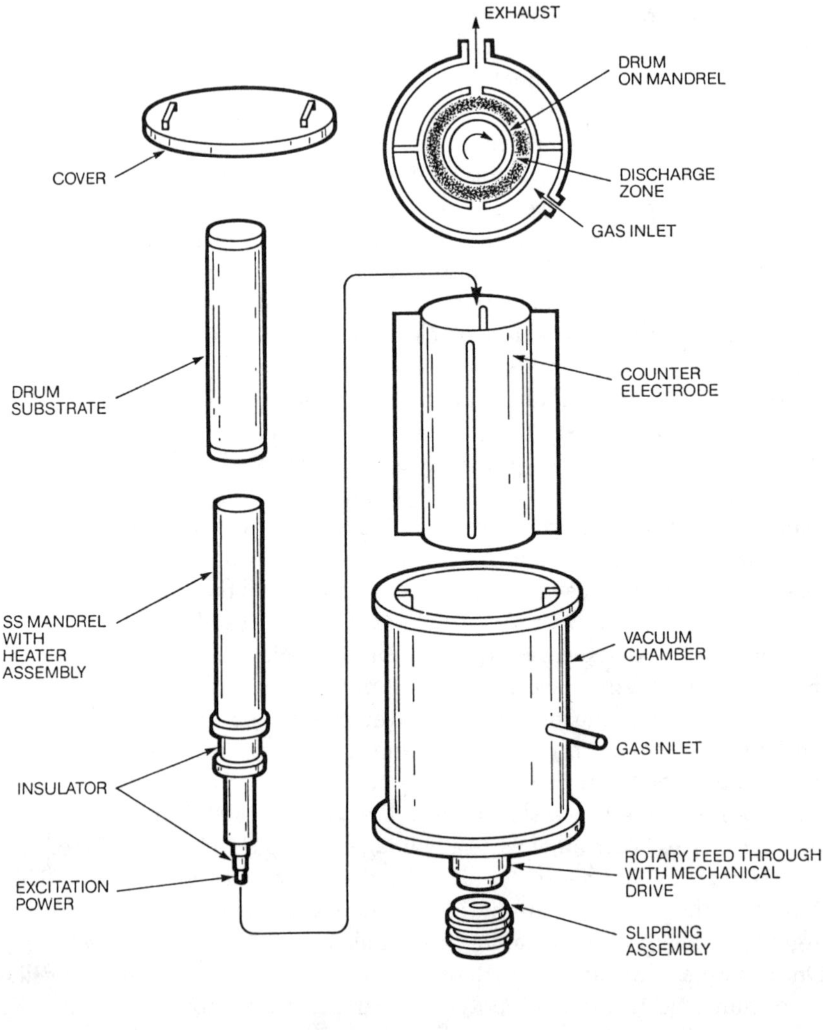

Hydrogenated amorphous silicon photoreceptors are unique in that they are fabricated from gases. The process involves the fragmentation of gases such as silane by the use of electric discharges first studied in the nineteenth century. Schematically shown is a reactor design for the production of amorphous silicon photoreceptor drums by the glow discharge of silane gas; the glow discharge is confined between the drum substrate and the counterelectrode. The primary advantage of amorphous silicon is its mechanical hardness which results in a photoreceptor of unusual durability. The material is also widely used as a photovoltaic generator of electricity since photovoltaic cells and xerographic photoreceptors share some of the same requirements.

the output power is the sum of the photogenerated power produced over the total area of the device. Thus although uniformity of properties and photoresponse over the total area are preferred, they are not in principle mandatory. Such is not the case with an image transducer such as a xerographic photoreceptor where excellent uniformity of properties over large areas is a sine non qua. The exploration of amorphous silicon for potential application as a photoreceptor therefore was not a simple extrapolation, and as it turned out a number of surprises were in store.

At this point one might ask why amorphous silicon photoreceptors could be important. Xerography already had chalcogenide and polymeric-based photoreceptors, each of which in its own way performed well. The new attribute that amorphous silicon offers in comparison is longevity. Unlike the other materials, amorphous silicon although clearly amorphous is not a glass. In fact it is a very hard, durable material and intuitively one might expect a substantially longer life resulting from a superior resistance to wear and impact damage.[116] Extensive research in the United States and Japan on amorphous silicon photoreceptors has been undertaken in the last few years.[117] The results of these endeavors show that the excellent photoelectronic properties observed in 1 micron films are maintained in films as thick as 40 microns. Both electrons and holes have carrier ranges of about 10^{-6} cm^2/V and the electron and hole mobilities are approximately 0.5 and 10^{-3} cm^2/Vs respectively. For use with hole transport, the silicon has to be lightly doped with boron at a level of ten parts per million (10 ppm). Amorphous silicon, however, was found to be unstable and reactive to the species generated by the corotron devices. This problem is solved by overcoating the silicon with a thin, 0.3 micron, passivation layer of glow discharge deposited silicon nitride (SiN_x) or silicon carbide (SiC_x). Fortunately both of these materials are equivalent or harder than amorphous silicon so the advantages of hardness are not compromised. All of these layers are easily produced by the glow discharge process and the total device can be produced in a single pump-down. Amorphous silicon photoreceptors which are now used in commerical photocopiers have a useful photosensitivity out to about 7500 Å. Current research is aimed at extending this spectral response further to solid-state laser wavelengths. One approach is to produce silicon-germanium alloys by the glow discharge of silane and germane (GeH_4) mixtures.[118] These materials have absorption edges which lie at longer wavelengths and therefore photosensitivity at the wavelengths emitted by solid-state lasers.

Progress in science and technology although occasionally marked by major or discontinuous breakthroughs is more typically incremental and evolutionary and achieved as the by-product of sustained, painstaking efforts. Consequently the advances made are most clearly discernable in hindsight. Such is the case for photoreceptor materials. The distance be-

A variety of amorphous silicon photoreceptor drums produced by plasma deposition. The 25 micron thick amorphous silicon layer is covered with 3000 Å passivation layer of silicon nitride, also plasma-deposited; films on planar substrates are also shown. A sense of scale is given by the circular disc, left foreground, which is a three inch single crystal silicon wafer. (Photograph, Courtesy Xerox Corporation)

tween Carlson's molten layers of sulfur or anthracene and the range, quality and capability of today's photoreceptors must be measured both in terms of effort expended and insight gained. As we have seen, this involved the development of the science of amorphous materials in the context of the solid-state band theory. If amorphous selenium spawned this new science then, just as the transistor was the by-product of the science of crystalline solids, amorphous silicon devices and organic photoreceptors involving molecularly-doped polymers were its fruits.

Chapter 4
Developer Materials and Processes: Now There's the Rub

Much of modern technology is like a chain, in which a number of interdependent elements must work in concert, since the failure of one produces failure of the whole. Such is the case for the technology of xerography. Vital links in that chain are the processes whereby the electrostatic image is rendered visible with toner, transferred to paper and fixed to create the final copy.[45,50,51] The processes while being distinct have one material in common, the toner. The development process depends on the action of the electrical fields created by the electrostatic charges at the photoreceptor surface on the toner. The toner consists of polymer particles approximately 10 microns in size within which much smaller carbon-black or other colored pigment particles are dispersed. The overall toner size is dictated by the degree of image resolution to be achieved by the development process but is so small that a single toner particle is undetectable by the unaided eye. Reverse processes are involved in the transfer step where the toner is transferred from the photoreceptor to corona-charged plain paper. In the fixing step the transferred toner is bonded to the paper by softening; most commonly by thermal melting. Thus as with the photoreceptor, the material requirements for the developer are diverse and demanding, although the phenomena involved are totally different. The toner must be manufacturable with both control and uniformity of its microscopic size and degree of charging and also have sufficient mechanical strength to withstand constant impaction and abrasion against the photoreceptor and other materials in the development system. In combination with all these properties, the toner has to soften with a minimum of stimulus from the fixing process and yet not clump in the xerographic machine when standing unused for long periods of time under a wide range of ambient temperatures and humidities.

Very few of these requirements were met in the original feasibility demonstration by Carlson.[44,45] the simple dusting of nonpigmented lycopodium powder over the exposed sulfur film yielded low density images of poor resolution with considerable powder in the exposed background areas, even after blowing to remove the excess. Such powder, in addition, did not have the capability of being transferred or fixed to paper without the latter having been previously coated with wax or other sticky material.

In this context, therefore, the development of organic polymer materials starting in the 1930s assumes critical importance in the success and continued development of xerographic technology.[119] Certainly, the early pioneers in polymer chemistry could have hardly dreamt of this particular application of their labors, and yet, without the ability to control mechanical, electrostatic and rheological properties, the combination of the required properties would not have been available when needed. These material advances, together with significant inventions by scientists at the Battelle and Xerox laboratories, in the control of triboelectricity, the delivery of the toner to the photoreceptor surface, its transfer to uncoated paper and subsequent fusing were no less significant than the development of the selenium photoreceptor in making Carlson's dream a reality.

The physical basis of development is simple in principle but complex in practice.[45,50,51] A uniform surface charge on a photoreceptor of infinite extent produces associated electric fields whose lines of force are confined within the photoreceptor since they terminate on the induced charges of opposite sign in its conductive substrate. In other words the external electric field is zero. This is not true if the photoreceptor has finite dimensions or there is a nonuniform distribution of surface charge, since discontinuities exist in the surface charge. At these edges the lines of force extend out of the photoreceptor into space. These so-called fringe fields, which are quite strong, are confined, however, to a narrow region at such a boundary. Thus a nonuniform distribution of surface charge produced by the imagewise exposure of a photoreceptor will also result in a fringe-field pattern across its surface. Fields external to the photoreceptor constitute the development field E_D which is literally the driving force because a toner particle of the appropriate polarity with a total charge Q will experience a force $F_D = Q \times E_D$ pulling it to the photoreceptor.

A major shortcoming of this fringe-field development is the fact that only narrow lines or edges of large-area black images can be developed, whereas the low development fields at the center of such solid areas lead to inadequate toning. This effect is clearly visible in the very first xerographic image produced by Carlson and Kornei.[45] Solution of this problem can be achieved by the use of a biased-development electrode placed close to the photoreceptor surface. In this way the lines of force are pulled out of the photoreceptor, even if it is uniformly charged. The development fields which fall off rapidly over distances of microns as one moves away from the photoreceptor surface exert an attractive force on a

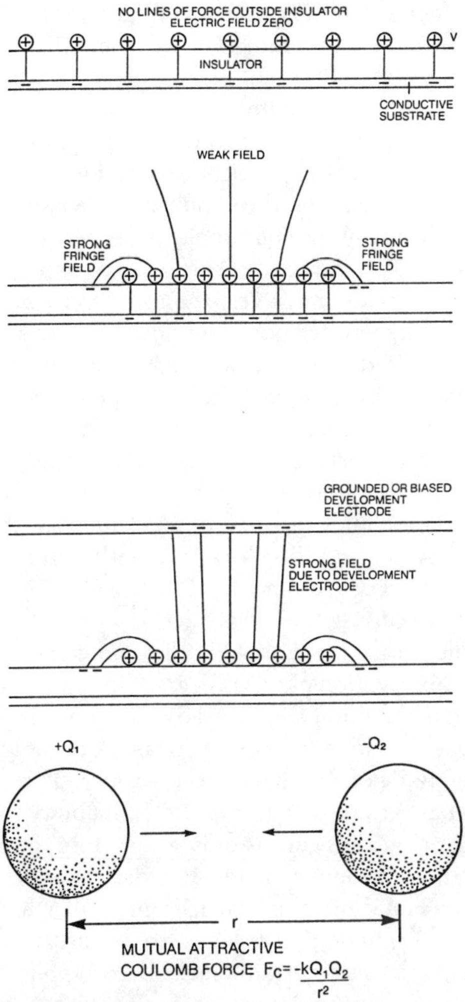

Top: *All the lines of force, and hence the electric field, are contained within a uniformly charged insulator of infinite extent. On the other hand, the lines of force fringe out above the photoreceptor for a nonuniform charge pattern produced by selective photodischarge, as in xerography. The field on charged toner particles is strongest at the edges of the charge distribution and constitutes the so-called fringe fields; the development field is much weaker in the center of broad-area images. This fringe-field development was operative in Carlson's first image. If a grounded or biased electrode is placed above the charged photoreceptor surface, the field strength in the center of broad image is enhanced. Such electrode development is used to produce uniform toner development of solid image areas.* Bottom: *The electrostatic or coulomb force is proportional to the product of the magnitudes of the two charges and, like the gravitational force, is inversely proportional to the distance separating two charges.*

triboelectrically charged particle of these dimensions. The average toner particle weighs about 1 nanogram (1 nanogram $=10^{-9}$ gram) so that in one ounce there are roughly 3×10^{10} toner particles; about ten times the world's population. One xerographic copy of a page of text can contain over 100 million toner particles, with about 100 in a single period. Thus the challenge is to exert control over such large numbers of particles so that they go where they are supposed to and nowhere else. Such discipline is achieved through the use of the coulombic forces between charged particles and the charges on the photoreceptor surface. To come under the influence of these electrostatic fields, however, the toner must be brought in close proximity to the charge pattern on the photoreceptor surface.

A variety of methods can be employed to achieve this delivery, most of which involve an intermediary component called a carrier. These different development processes include powder-cloud, cascade and magnetic-brush development.[45,50,5] Liquid development in which an insulating hydrocarbon liquid is used as the carrier will not be discussed further except to say that it allows the use of much finer toner than can be controlled as a dry powder and therefore inherently offers the possibility of much higher resolution development.[20] Powder-cloud or aerosol development used air as the carrier.[121] The attraction of the micron-sized carbon particles of candle smoke by charged bodies, noted by Gilbert in 1600, is the first recorded observation of this process.[45] In its basic elements this was also the development method employed by Carlson. The major difficulties with powder-cloud development arise from a lack of sufficient control over the sign and magnitude of the charge on the toner. This typically leads to excessively dark backgrounds. There is also an inability to manage the majority of the toner cloud, since it is only toner close to the photoreceptor that experiences the development fields. Powder-cloud development, as the name implies, is a cloud of toner particles most of which respond to air movement so that inherently it leads to a dirty environment in a xerographic machine. For certain specialized applications, however, aerosol development can be the process of choice. One example is in xeroradiography where small changes in transmitted x-ray intensity need to be detected. Without a development electrode, the xerographic images give much higher image contrast than conventional radiology thus enhancing discontinuities such as fractures. In other cases the use of a development electrode can provide high resolution, continuous tone xeroradiographic images. Xeroradiography shows particular benefits in mammography and provides an exceptionally sensitive screening method for breast cancer.[50]

Some processes utilize solid carriers for toner delivery, so that the developer consists of both toner and carrier, and are known as two-component development.[122] Typically the carrier constitutes over 95 percent of the total weight of the developer and, unlike the toner, is not

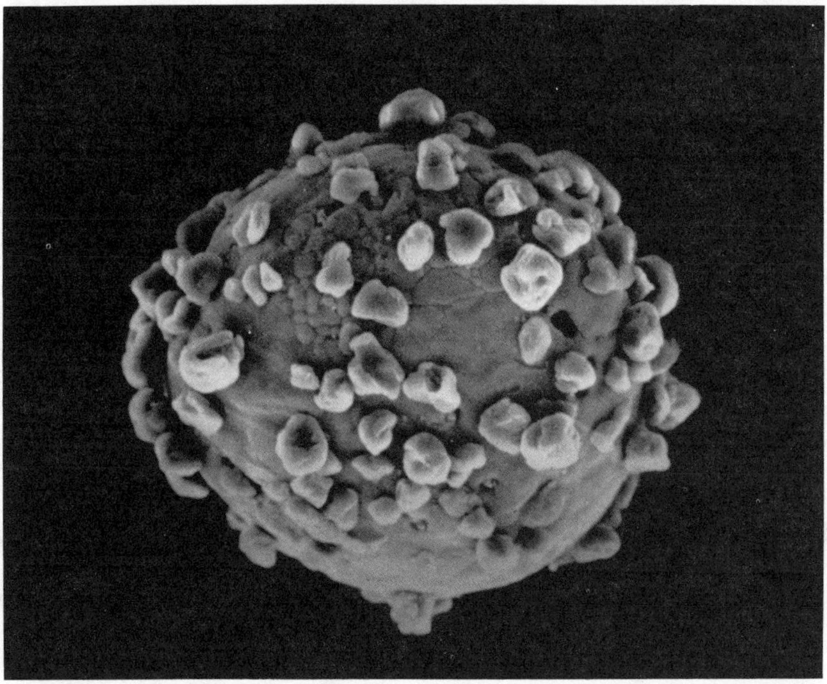

Magnified photograph of a single carrier bead 100 microns in diameter showing attached toner particles which are a few microns in diameter. (Photograph, Courtesy Xerox Corporation)

intended to be consumed in use. The first process called cascade development is characterized by a dependence on gravitational forces for its operation. In the second process, magnetic-brush development, the carrier is made of a magnetic material such as iron so that magnetic forces supersede gravity as the dominant delivery force. Magnetic-brush development was first described by RCA in 1953 and is now the predominant development process used in the copier business.[123,124] A single carrier particle or bead is much larger in size than the toner and in early products had an average size of about 700 microns, although today 100 microns is more representative. The carrier not only serves to physically deliver the toner to the surface of the photoreceptor but also to control and determine its charge. The carrier particles which are roughly spherical have an average weight of a few millionths of a gram and are often coated with a thin insulating polymeric shell which controls the degree and sign of charge imparted to the toner. An additional requirement is that this coating provide sufficient adhesion so the toner particles can piggy-back on the carrier, although this adhesive force cannot be so large that adhesion is permanent. A very delicate balance is therefore required between the various possible competing electrostatic, adhesive, magnetic and gravitational forces.

Although there are specific differences between these and other development systems, there are issues relating to the phenomena of charge control that can be discussed generically. For development to occur, the toner particles must be charged with reproducibility to the correct level and polarity for the specific photoreceptor. The phenomenon of triboelectricity, first described by the Greeks, is employed.[45,125,126] The understanding of this phenomenon is by no means complete in the sense that no absolute correlations between theoretical predictions and experimental observations can be said to exist. Despite this, a considerable phenomenological and operational understanding together with control of development materials have been established. The problems in achieving a full descriptive and predictive understanding of triboelectricity arise because the phenomena are associated with the surfaces of solids, their interaction with each other and their ambient.[125] The specific experimental observations made can be sensitive to the nature and purity of the materials, the physical and chemical state of both surfaces and the precise details of the experiments.

The fact that the name triboelectricity comes from the Greek word "to rub" suggests that rubbing is essential to the mutual charging of two solids; a conclusion that might also be deduced from its original discovery. However, it is possible that rubbing only serves to provide an area of contact between two solids and that this is the true mechanism of charge transfer. However, the relative roles of rubbing and contact are not easily determined. The achievement of intimate contact between two solids is extraordinarily difficult since particulates like dust on a surface are almost impossible to avoid and remove. Thus with particle sizes of a few microns, it is hard to bring two optical flats into intimate contact and typically this can only be achieved by pressure and twisting. It is difficult to decide under such circumstances whether the observed charging is directly due to the frictional rubbing or whether this merely enhances the contact by increasing the number of contact points. Such considerations are always complicating factors in attempts to deconvolute the relative roles of contact or rubbing by controlled experiments. Shaw showed, for example, as early as 1917, that the nature of triboelectric charging between the same two materials was strongly dependent on the nature of the rubbing.[127] Thus if quartz was gently brushed with silk it acquired a negative charge, whereas if hard rubbing was involved it became positively charged. Today, however, it is believed that charging can occur by contact alone, as demonstrated by experiments where one of the materials is liquid mercury. It is likely, therefore, that both contact and frictional rubbing can be important. Whatever the precise mechanism, it is clear that triboelectricity involves the movement of charge from one solid to another.[125]

POSITIVE

Window glass

Polyamide

Common salt

Wool

Silk

Cotton

Styrene butadiene copolymer

Polyvinyl butyral

Epoxide resin

Natural rubber

Sulfur

Polyethylene

Polyvinylchloride

Polytetrafluoroethylene

NEGATIVE

TRIBOELECTRIC SERIES OF VARIOUS INSULATORS

A triboelectric series in which the relative charging properties of materials are listed. Materials at the top of the list charge positively with respect to those at the bottom. Although variability due to dependence on measurement conditions precludes the establishment of a quantitative scale, such a series serves as a valuable guide to the materials and xerographic scientist.

Despite the complex variety of possibilities as to how triboelectrification occurs and the inevitable variations in experimental observations, a correlation between material properties and their triboelectric behavior has often been empirically perceived, albeit dimly. Inherently triboelectricity is a relative phenomena since at least two different materials are involved. As a result much use is made of ordering the relative charging capabilities of materials. Constructed from empirical studies conducted under standardized procedures, this is known as a triboelectric series.[125,128] Materials at the top of the series charge positively with respect to materials near the bottom. Such series are used only as general guides and the results of any particular combination may deviate from expectation based on possible contamination, impurities or humidity variations. A number of efforts have been made to correlate such series with other material parameters. Since the charge exchange may involve the transfer of ions, electrons, or macroscopic amounts of charged material, either separately or in combination, it is perhaps not surprising that a number of such relationships have been reported. In the mid-nineteenth century, Faraday studied triboelectric phenomena in a number of liquid and gaseous substances and established a table according to their electrification abilities.[14] In 1898, Coehn postulated a law bearing his name stating that materials of higher dielectric constant become positively charged when contacted with those of lower dielectric constant.[129] A correlation with dielectric constant is consistent with a charge transfer of weakly bonded ions between solids, since the relative binding energies of these ions to their respective surfaces are affected by screening, through the dielectric constant. Another correlation involves the work function of materials which is a measure of the energy required to remove an electron from a material. This is based on a picture in which the charge transfer requires the movement of electrons from one material to another and is related to the relative difference in electron energy levels in the two materials.

To gain some perspective on the development process, it is useful to quantify some of the parameters. For this purpose, consider a photoreceptor 50 microns thick, charged uniformly to a surface potential of 1000 volts. With a typical dielectric constant of 5, the amount of charge on each square centimeter of the photoreceptor is 100 nanocoulombs. Since one electronic charge amounts to 1.6×10^{-19} coulombs, each square centimeter of surface would be occupied by 6×10^{11} ions if this surface charge were due to singly charged ions. On exposure of the photoreceptor to an image pattern of variable light intensity, nonuniform photodischarge occurs producing a latent electrostatic image and related development fields which replicates the information content of the document. To achieve complete photodischarge, 6×10^{11} carriers must be generated for each square centimeter and with a photogeneration efficiency of unity this requires an exposure of less than 10 ergs/cm^2; this is one millionth of the energy needed to lift one pound in weight to a height of one inch![74]

Now the photoreceptor is always charged to a particular polarity using the appropriate sign voltage in the corotron, so the toner particles must be charged to the opposite polarity in order to produce a black on white copy of a black on white original. If the input were a white on black, i.e., a

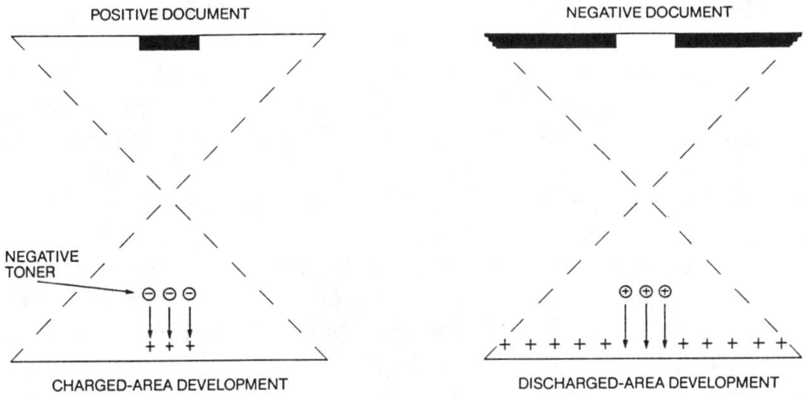

Charged-area development uses the toner charged with opposite polarity to that used to charge the photoreceptor; discharged-area development uses the same sign toner. In this way whether a positive, black text on white, or a negative, white text on black, original, is used a positive copy will be produced.

negative, then toner with the same polarity would be required to give a black on white copy. This is an example of reverse or discharged-area development.[130] Photoreceptors in different products are also used with either positive or negative charging, so appropriately charged toner of either sign must be available. The charging in both sign and magnitude is controlled by the nature of the toner and carrier which in turn may have a special coating. The degree of triboelectric charging and its uniformity is important to achieve the most efficient development and highest quality images. The blackness or density of the image is related to the amount of toner which deposits on the photoreceptor, so the charge to mass ratio for the toner is an important parameter. The toner/carrier interactions lead to a mutual charging of from 5–20 nanocoulombs per square centimeter. This corresponds to about 10^5 electronic units of charge per toner particle or a charge to mass ratio of 10 microcoulombs per gram of toner. For a typical toner, a deposition of about 1 milligram of toner per square centimeter will give the dark image representative of a high quality copy. If the charge to mass ratio is 10 microcoulombs/gram, then the amount of toner needed to give a sufficiently dark image will neutralize 10 nanocoulombs per square centimeter. Since this is only about one tenth of the charge density on the photoreceptor, sufficient development can take place without substantially reducing the development field.[130]

Although the actual fabrication processes of the developer materials will be discussed later, it is sufficient for now to know that the very small toner particles are produced by a grinding or milling process. This involves the continual fracturing of the polymeric materials so that charges of both polarities are created. Thus the toner material itself after it is fabricated, although uncharged on the average, consists of quantities of positive and negative charge. The number of particles with a given charge or mass can be measured by a charge spectrometer in manner not very different to that used by J.J. Thomson in his discovery of the electron.[20] Such experiments on freshly made toner show that the charge distribution is indeed symmetrically distributed about zero charge, but that equal fractions of the toner particles have nonzero opposite charges.[131] This situation changes when the toner is mixed with carrier beads. The charge distribution now shifts so that the average charge is nonzero. The sign of this new equilibrium charge of the developer depends on the choice of the materials for the component toner and developer or any additives. The magnitude of the charge depends both on the time of mixing and the ratio of toner to carrier or toner concentration in the mix. With time, the average charge grows approximately in an inverse exponential manner and saturates to a value determined by the toner concentration.[132] The inverse exponential growth is consistent with a charging mechanism involving contact in which the fraction of uncharged area on a carrier bead decreases with time. This reflects the fact that the charge density per unit area of a toner particle depends on the probability of finding an uncharged area on the surface of a carrier. This becomes increasingly difficult as the toner concentration C_T, defined as the toner mass divided by the sum of the toner and carrier masses, increases. Based on this intuitive picture, the charge to mass ratio of the toner is expected and found to decrease with increasing C_T. Thus it would seem that the engineer would merely have to design the material package to obtain a developer with the right polarity of triboelectric charge and toner concentration to give the desired charge/mass ratio. All that would be required would be to maintain the toner concentration by adding toner as it was depleted. Although this is true in principle, other considerations arise.

The developer materials are subjected to considerable electrodynamic forces amounting to several thousand g's, particularly in high speed magnetic-brush development systems. As a consequence abrasion of carrier coatings, fracturing of carrier beads and compression of toner particles can and do occur.[51] Since carrier beads are commonly coated with low surface energy polymer coatings for charge control and minimization of adhesive forces, such events can have serious deleterious effects. Exposure of the underlying carrier beads which may have relatively high surface energies means that any toner compressed onto that surface can become

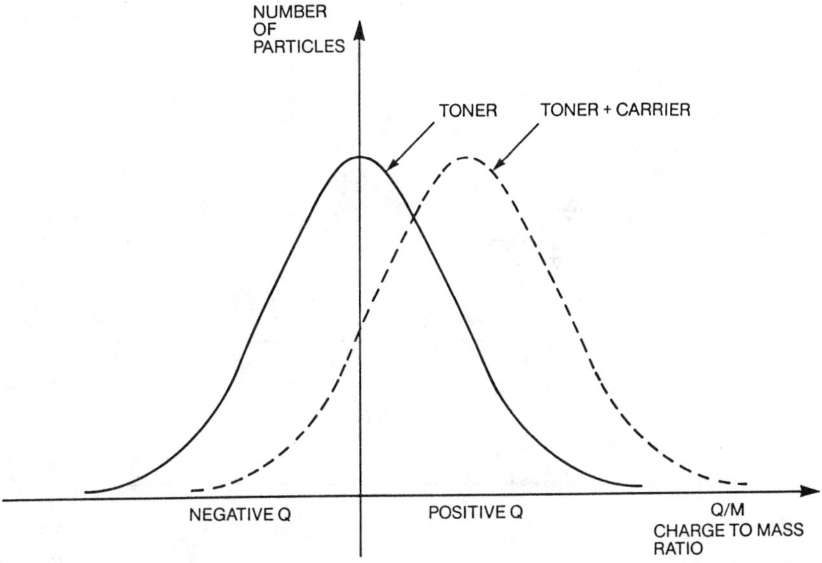

Individual toner particles produced by milling are initially charged either positively or negatively. Collectively, on the average, the net charge of toner is zero. On mixing with carrier particles, however, the symmetric distribution about zero charge to mass ratio shifts and the toner develops a net positive or negative charge depending on the nature of the materials involved. The degree of shift also depends on the time of admixing since it is determined by the integrated number of contact charging events.

permanently bonded to it. The net effect is that part of the carrier surface now presents a toner surface to the other toner particles so that on the average no charge exchange can take place at those points. This is one example of the many aging phenomena that can occur with developer materials. Such aging phenomena reveal themselves in progressive loss of image quality and ultimately require the replacement of the total developer package.

Whatever the mechanism of charging, viewed from the photoreceptor, charged toner particles attached to carrier beads come into close proximity with its surface. Seen from the toner's perspective the situation is somewhat different. The toner/carrier system approaches the charged photoreceptor with the toner bound to the carrier with an adhesive force. This adhesive force has two components, an electrostatic force F_E and a dispersion force F_A and their net effect must be overcome by the development force exerted on the toner by the latent charge pattern on the photoreceptor.[133,134] Since

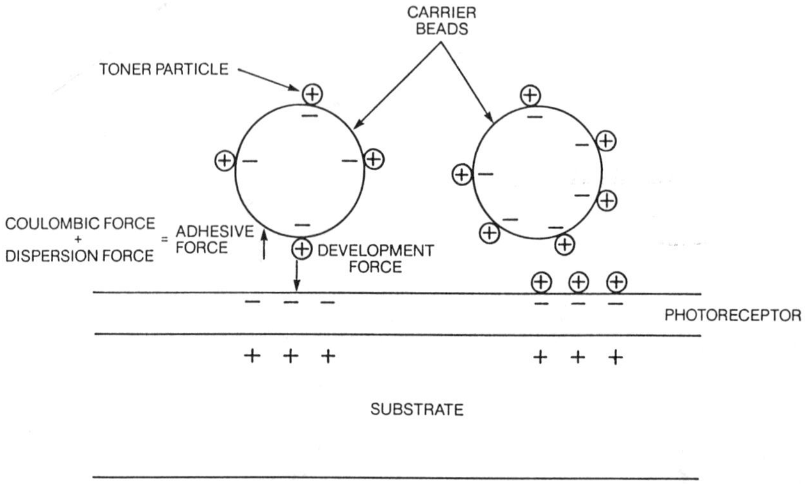

Two-component development involves a competition between electrostatic and adhesive forces. The adhesive force, which is a combination of the electrostatic force between the charged toner and the oppositely charged carrier bead and the dispersion force, must be overcome by the development force produced on the toner by the photoreceptor surface charge distribution. A variety of collisions occur both between the developer materials themselves and with the photoreceptor surface. These can be beneficial by inducing sufficient separation of the toner from the carrier to decrease the highly distance-dependent dispersion force and thus aid the developability. However, such impacts can also lead to aging of developer materials due to permanent impaction of toner onto carrier beads and the loss of charging efficiency.

the toner acquired its charge from the carrier, the carrier has a charge of opposite sign and the pair will be physically bound by their mutual electrostatic attraction. Experimentally it is known that typically a single toner particle carries a charge of about 10^{-14} coulombs (equivalent to about 10^5 electrons). The electrostatic force, F_E, binding a spherically shaped toner/carrier pair is proportional to Q^2/r^2, where Q is the assumed uniformly distributed charge on both the toner and the carrier and r is the radius of the toner. F_E is of the order 10^{-8} Newtons so the electrostatic force on a toner particle is extremely small, since a Newton is the force that gives a mass of one kilogram an acceleration of about 2 miles per hour/sec. The dispersion force, F_A, is a form of Van der Waals bonding associated with the interaction of fluctuating dipoles generated by otherwise neutral atoms.[51,135,136] The dependence of the size of this force on the separation of the two interacting atoms is distinctly different from the inverse square relationship of the electrostatic force between two charges. It takes a very short, but finite, time for the electromagnetic signal from one fluctuating dipole to reach a second. In response to the arrival of this signal, the second atom polarizes and emits its own dipole radiation. If the electromagnetic signal from the second atom is felt back at the first atom before the latter changes its original dipole configuration, then the interactive force is said to be in phase and is attractive in nature. The requirement is that the total time involved, which is the round trip transit of an electromagnetic signal between the two separated atoms, be less than the period of oscillation of the fluctuating atom. For atomic separations less than about 100 Å, this condition can be met and an order of magnitude estimate of the dispersion force, F_A, between a typical toner/carrier pair gives a value comparable or higher than the electrostatic adhesive force. For distances substantially greater than 100 Å, any phase correlation is lost and the attractive nature of the dispersion force decreases. Thus the overall strength of the adhesive force can be substantially reduced if sufficient agitation in the development process can detach and separate, even slightly, the toner particles from the carrier beads. This then enables the development field to overcome the adhesive force binding the toner to the carrier.[51]

To summarize, development depends on whether the development force F_D exceeds the forces F_A+F_E keeping the toner attached to the carrier. The dispersion force F_A does not depend on the charge on the toner and for a given set of materials is constant. The electrostatic adhesive force, on the other hand, is proportional to Q^2/r^2. Hence for a given development field the charge on the toner may be so small that the development force QE_D is less than F_A+F_E and development will not happen. As the charge on the toner is increased, however, and/or the toner radius is made smaller then F_E grows faster than the development force and again development will not occur. This means there must be an optimum range of values for Q^2/r^2 and therefore for Q/M, the charge to mass ratio, for the toner.

In cascade development the gravitational force on the much more massive carrier particle is used to cascade it, together with its attached toner, over the photoreceptor surface. The carrier particles roll and tumble across the photoreceptor surface so that two effects are important. In bouncing the physical impact shakes loose some of the toner. As a result it becomes airborne and since the physical impact has already supplied some of the stripping energy, the airborne particles develop out at relatively low development fields. Toner particles remaining bound to the carrier beads require higher fields for development. Within a few microns of the surface, this system of toner and carrier feels the effects of the development fields which for a charge of 10^{-14} coulombs pull with a force of 10^{-8}

Newtons. In the world of a single toner particle with a mass of only about 10^{-9} grams, however, the effect of this small force is such that, in the absence of other forces, the toner particle would be given an acceleration of 200 miles per hour/sec! Thus this force ultimately wins and the toner particle moves towards and becomes attached to the photoreceptor. The retention forces between the toner and carrier in two-component developers give such systems a higher contrast because of the threshold effect since, unless the development fields exceed the forces retaining the toner on the carrier, no deposition occurs.[130] This contributes to clean, dust-free background on nonimage areas, the absence of which plagued Carlson's original invention. On detachment from the carrier, the toner in addition to the pull from the development field, also experiences the gravitational force, although for typical toner sizes this force is less than 0.1 percent of the development force and is negligible.

Many materials were investigated by Battelle as possible toners and carrier materials. Perhaps as a testament to the pervasiveness of tribo-electrical phenomena, one of the materials studied, successfully, as a carrier was corn meal.[137] The earliest copiers used an insulating carrier of sand and had a major shortcoming in their inability to reproduce solid black-areas because development was only by fringe fields. Improvements were achieved using a closely spaced development electrode or conductive steel shot as the carrier, since a conductive carrier can serve as a development electrode. Another difficulty with cascade development was the active role of gravitational force. For a given carrier bead material, the gravitational force which equals the carrier weight is proportional to the cube of the bead radius, whereas the carrier charge, related to the charge per unit area, is proportional to the radius squared. Carrier particles, unfortunately, can have a distribution in sizes so that small carrier particles can be culled out of the cascade flow and become attached to the photoreceptor because the electrostatic forces on them can exceed those due to gravity. In addition to possible physical damage to the photoreceptor surface, this limited the achievable flow in cascade development and was a constraint for higher machine speeds.[51] This problem was overcome by making the carriers out of magnetic materials so that, by using magnets, the magnetic fields could dominate over gravity. Most modern xerographic machines now use magnetic carriers and magnetic-brush development.

In magnetic-brush development the carrier is made of very small (100 microns) beads of ferromagnetic material which are coated with a thin polymer charge-control film.[123] Spherical carrier beads have the advantage of flowing more freely so that fresh toner, which must be added from time to time, is charged faster and with greater uniformity. Also the absence of sharp asperities prolongs developer life by allowing more uniform surface

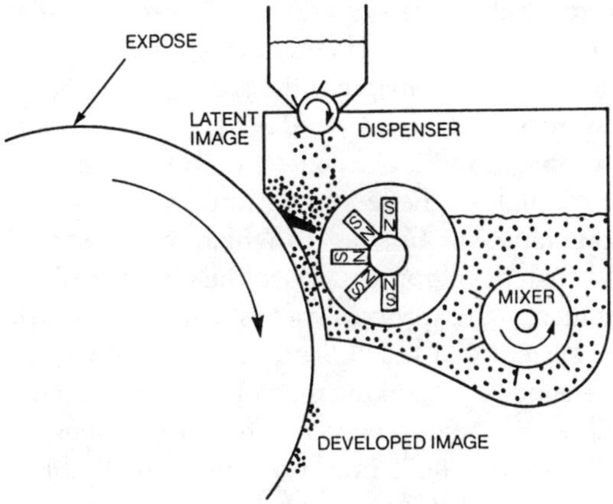

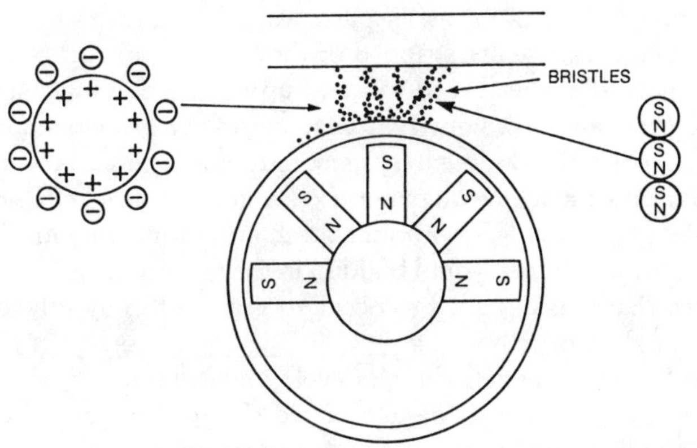

Two-component magnetic brush development involves the use of magnetic carrier beads which carry nonmagnetic toner particles. The transport of the carrier and attached toner is accomplished by a rotating cylinder enclosing an assembly of permanent magnets which provide the magnetic force that prevails over centrifugal and gravitational forces. Within the magnetic fields, the magnetic carrier particles behave as individual bar magnets that through mutual attraction and repulsion of their magnetic poles form chains or "bristles."

coatings and minimizing the attrition caused by impacts. Ferrite carriers are favored because they are about 50 percent lighter than steel and have a much lower remanent magnetic field, which means that in the absence of an external magnetic field they retain less magnetism. This causes them to mix more readily and flow better in the developer reservoir which is uninfluenced by the developer-roll magnets. The latter, which provide the necessary magnetic fields, are stationary permanent magnets located within a rotating drum or sleeve which transports the toner-carrier combination to the photoreceptor surface.[124] The magnetic fields cause the magnetic cores of the carriers to become individual magnets with north and south poles. As a result of their mutual magnetic interaction, the carrier particles link up and align to form chains or "bristles" to which toner is attached, hence the name magnetic-brush development. With the elimination of gravity as a dominant force, much higher process speeds can be achieved. Velocities of the carrier-toner system can reach 5 miles per hour which, given the relatively small dimensions of a copying machine, means that the developer mix is recirculated every few seconds. The bristle length, which is something of a random variable, is important in magnetic-brush development since it determines the degree of the toner contact with the photoreceptor and to some degree the quality of the developed image. The relative speed of the rotating developer drum with respect to the moving photoreceptor also plays a role. At too high a speed, however, the centrifugal force can result in carrier particles escaping from the constraining magnetic field. If these are stripped of the toner by impacts, then the oppositely charged carrier beads can end up on the nonimage areas of the photoreceptor; an effect known as bead carry-out.[51] Incidentally as toner moves from the development system onto the photoreceptor and ultimately out of the machine on paper, excess unneutralized charge can build up on the carrier particles. Without a mechanism for eliminating this, some very unhealthy voltages would buildup in the developer unit. Discharge of this excess charge is designed to occur to ground through the conductive walls of the developer housing.

Two types of magnetic-brush development, known as insulating and conductive magnetic brush, are employed. The distinction is based on the effective conductivity of the developer and more specifically the assembly of contacting carrier particles.[138] This is controlled by the chemical composition and thickness of the polymer coating of the carrier beads. Conductive magnetic brush gives improved solid-area development because of the reduced effective spacing between the photoreceptor and the developer roller caused by the conductive paths within the developer. In other words the developer itself functions as a development electrode. Within the turbulent maelstrom of the rotating magnetic brush, a whole series of damaging phenomena can and do occur including collisions between carrier and toner particles and the photoreceptor surface. These can lead to the beneficial impact-associated separation between the toner and the carrier discussed earlier. But physical impaction and permanent bonding of toner particles to the carrier surface or fracture of the toner or carrier can also occur, as was the case with two component nonmagnetic brush development. All of these contribute to the aging of the developer material because they change the charge to mass ratio that any new toner, periodically added to replenish that used up in producing copies, can attain. Such aging can lead to a diminishing image quality in terms of contrast, background and loss of resolution.[51]

A more recent change in development systems is the use of single or mono-component developers.[139] As the name implies, only one kind of particle is used. This serves as the toner and can be magnetic or nonmagnetic depending on whether ferromagnetic material is incorporated. One drawback of magnetic toners is when colored developers are needed, since materials with the required magnetism are typically brown to black in color and limit the quality of colors achievable. No carrier particles are used, so that in this respect the system is simpler, smaller and avoids some of the aging effects just discussed. There is another material, however, involved in order to control the necessary triboelectric charging of the toner. The charge control is accomplished by coating a roller with the appropriate material, just as carriers are coated in two-component developers. This roller thus acts not only to control the charging but also the transport and metering of the toner. This, or a similar roller, can also be appropriately biased to function as a development electrode. A particular case of single-component development is jumping-toner development in which triboelectrically charged toner covering a donor roll rotates synchronously with the moving photoreceptor. By use of electric fields between the donor roll and photoreceptor, selective transfer of toner occurs. The simplicity, conciseness and low cost of this approach enabled the introduction of the throwaway cartridge concept for copiers.

Once the latent electrostatic image has been developed on the photoreceptor surface, it must be transferred with no loss in clarity.[140] The identification of the electrostatic transfer process at Battelle was yet another invention crucial to the commercialization of the xerographic process. The concept was to bring a plain, uncoated sheet of paper into contact with the photoreceptor without smudging. This was then followed by corona charging the paper with the same polarity of charge (assumed positive) to that used to originally charge the photoreceptor. Prior to the corona charging, the paper actually will seldom make intimate contact with the photoreceptor except at a few points; elsewhere an airgap separation of several microns is likely to exist. On charging with positive ions, negative charge is induced in the substrate and the top surface of the photoreceptor. Since, except under extremely high humidity, paper is a reasonably good insulator it consequently acquires a potential of several hundred volts during this process. As a result in unexposed areas an attractive field exceeding 10^5 V/cm exists between the paper and the photoreceptor. The pressure from this force pushes the paper into intimate contact with the photoreceptor

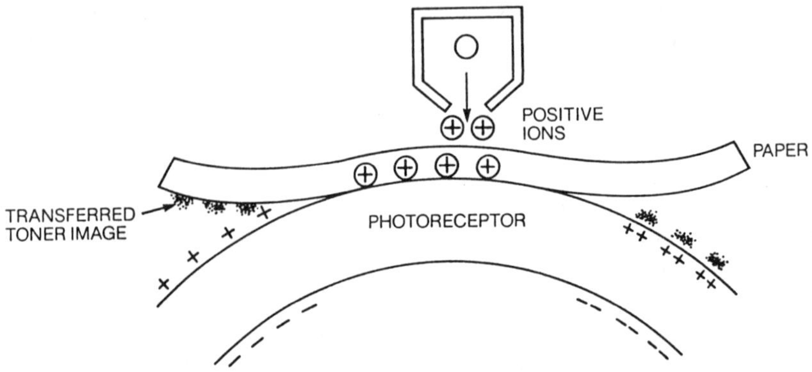

After the latent electrostatic image has been developed on the photoreceptor surface, the toned image must be transferred to paper with essentially no loss of clarity or resolution and high efficiency. This is accomplished using a transfer corotron which charges the paper with the same polarity of ions used in originally charging the photoreceptor. In addition to attracting the oppositely charged toner from the photoreceptor, the electrostatic force produces intimate and uniform contact between the paper and photoreceptor.

and ensures that the toner adheres to the paper following separation; the transfer efficiency is maximized by the insulating character of the toner. On a typical page of text there are roughly 50 milligrams of toner and routinely 80 to 90 percent of this can be transferred, even at process speeds in excess of 2 copies per second. One interesting benefit of electrostatic transfer is that the transferred image is often sharper than the developed image on the photoreceptor. This occurs if there is any wrong sign toner in the developer. Although such wrong sign toner is negligible in fresh developer, it can increase as aging occurs. In the development process, such toner can be attracted to image edges. Here because the field is reversed, wrong sign toner is pushed into regions beyond the image edge and can produce a slight blurring. During the transfer step, however, this toner will be repelled by the strong reverse fields when the paper is electrostatically charged and a more faithful reproduction of the document results.

The discussion of developer materials from the perspective of the electrostatic and/or magnetic and thermal interactions involved in the image development, transfer and fixing, highlights the interdependence that must be maintained between a range of diverse material properties. Since both the triboelectric and adhesive phenomena are surface sensitive, one might expect that the fabrication of the carriers and toners would require a very delicate and precise manufacturing technique in terms of the purity of materials and cleanliness of the ambient. Now although a high degree of quality control is both necessary and employed, the physical aspects of the manufacturing process are anything but delicate. Nor is this a minor activity. Currently, thousands of tons of toner, carbon-black and carrier materials are produced and consumed each year with a retail value of several billion dollars. To give some sense of the practical issues involved in the choice of materials and their manufacture, the toner and carrier will be discussed in turn. The precise details of materials, process and procedures are closely held proprietary secrets of the copier companies, so the discussion will inevitably be general.[137] Nonetheless it will give some sense of the various approaches used. The emphasis will be on two-component magnetic developer materials since this is by far the most widely used development process.

Together with achieving and maintaining the important triboelectric characteristics, such as polarity and charge to mass ratio, many other critical material properties must simultaneously be considered. At the same time the material choices must be consistent with volume manufacturing, reproducibility and cost. The material must not be toxic, allergenic or release unacceptable odors or fumes on heating in the fusing step. They must be insensitive to moisture and not cake either in storage or standing in a machine under a wide range of ambient conditions. The yield of toner, i.e., how many copies can be made per pound of toner, is also an important consideration. This is a somewhat elusive parameter since the fraction of a page that an image takes up can range from a few percent for a few lines of text, to 30 to 50 percent if substantial solid areas such as pictures are involved. For a standardized type of use, nonetheless, toners of low density will give higher yields than heavier ones such as single-component developers where the pigment is a heavy magnetic particle. The flow properties of the toner in the machine are affected by both particle shape and size distribution but to date, the only way to manufacture the optimally-shaped spherical toner particles of the required size and distribution is solvent-spray drying and this method has two major drawbacks. The first is cost and the second is solvent retention which with time, as the trapped solvent migrates to the surface, causes clumping.

The most widely used manufacturing process for making toner, therefore depends on using brittle materials since these can be milled or pulverized, at a high rate; this increases production volume and reduces costs. The downside is that an exceptionally brittle material can result in the production of "fines" which are unusually small particles, say < 5

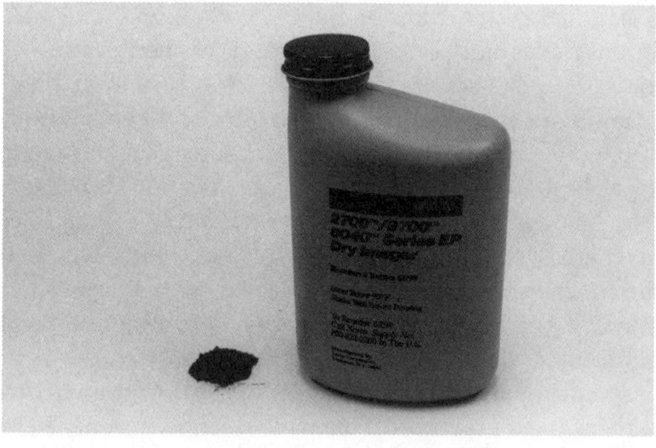

Top: *Bulk toner produced by mixing a polymer with carbon black or a colored pigment. At this stage, prior to pulverizing and jetting to produce the micron-size powder used in xerographic development, its taffy-like consistency is clearly apparent.* (Photograph, Courtesy Xerox Corporation) Bottom: *The pile of toner shown contains about a billion toner particles.* (Photograph, Courtesy Xerox Corporation)

microns. Excessive amounts of fines can reduce powder flow and cause background.[51] Finally the thermal properties of the toner polymer, which are critical in determining the fusing characteristics, must be considered together with negative temperature-related effects such as a tendency to clump.[51,140]

Almost all black toners consist of a thermoplastic polymer or polymer blend and carbon-black as the pigment in a weight ratio of about 85 percent polymer and 15 percent carbon-black. The earliest toner formulations utilized a polystyrene/carbon-black mixture. However the properties of the polystyrene led to unacceptable filming of the photoreceptor. This was solved by blending an acrylic resin, polybutylmethacrylate, with the polystyrene. Styrene, acrylic and polyester polymers and copolymers are largely used for negatively-charged toners for use with positively-charged photoreceptors such as amorphous selenium and its alloys, while polyamides, polyethylene and ethylene-vinyl acelate copolymers are used as positively-charged toners with polymeric-based photoreceptors which are typically used with negative charge. The carbon-black pigment is produced by the burning of oil in specially designed furnaces which produces the extremely fine particle sizes (a few tens of Å) required. For color xerography the three primary colors cyan, magenta and yellow are needed for process or continuous color images.[50] A variety of organic pigments exist that provide these necessary colors.[141] Color is, however, much more subjective than black or white so that the concentration of the added color calls for particular skill and judgment. The toner resin and pigment are mixed in either a batchwise or continuous operation using high shear mixers to produce a blend with a taffy-like consistency. This is rolled into slabs and cooled. The slabs are cut or broken into smaller pieces before being fed first into a crusher and then a pulverizer. The resultant powder or slurry is then jetted using high speed air-flow and the ensuing inter-particle collisions and abrasion produces toner-size powder. This powder is then screened to remove coarser particles and classified into particular size distribution by a number of techniques. In one of the final processing steps, additives such as metallic stearates or other materials to aid in the flow and cleaning properties may be added.[142] Finally the toner and developer are comingled so that the triboelectric properties such as charge per particle, charge per mass ratio etc., can be determined as a final qualification. It can be determined in this way that the total development package meets the specifications intended and required for the xerographic product to function as advertised.

The composition of carrier includes the core material and various carrier coatings that may be used and whether, as will be assumed, it is intended for use with magnetic-brush development. A variety of carrier-core materials have been used including spherical or nonspherical steel shot, sponge or porous iron and ferrites. In addition to their better flow and mixing properties, ferrites are already in a fully oxidized state so that their resistivity is stabilized.[143] A variety of processes are used to produce carrier materials, these include atomizing, electrolytic and chemical reduction methods to produce steel or iron carriers or spraying techniques for the ferrites. The carrier beads are then typically coated with a thin polymer film to provide a uniform carrier surface, which acts as a low surface energy layer and reduces toner filming of the carrier. Such carrier coatings also control the charge polarity and charge to mass ratio and prolong developer and photoreceptor life.

Although paper is clearly an indispensable part of the process of making copies, it is perhaps viewed as a passive element.[144] This far from the truth since it has to be reevaluated in terms of the varying phenomena with which it becomes implicated in the xerographic process.[51] Its electrostatic and electrical properties are important not only in the transfer process itself but also in its stripping from the photoreceptor surface. Other aspects of paper are also important in determining its motion at high speed through the machine and in the fixing or fusing step. Actually paper is a complex material both in terms of its structure and composition. In its basic form it is an interpenetrating network of cellulose fibers with a significant amount of free volume occupied by air. Xerographic-grade paper typically contains significant amount of fillers to enhance its mechanical strength, smoothness and reduce the shedding of fibers. All of these factors are important in minimizing the risk of jams as the paper makes its journey through the machine. The fillers also allow optimization of the optical properties including whiteness and reflectivity. This is obviously important in determining the contrast between an image character and the background of the paper. Paper is of particular concern in the transfer and fusing steps because of its propensity to absorb moisture. This is in part due to the contained air, arising from the porosity of the paper, but also because of the hydrophilic or water-attracting nature of cellulose. For example, paper in an ambient relative humidity of 50 percent will often contain up to 8 percent of moisture. This can lead to orders of magnitude increase in the conductivity of the paper.[51] Physically this results in the ions from the transfer corotron flowing through the paper onto the photoreceptor surface and can affect the efficiency of the transfer. While the use of heavier, less porous paper leads to higher transfer efficiencies, modern copiers are designed to handle a wide range of paper sizes and weights.

Having transferred the toner image from the photoreceptor to the paper, the image must be fixed or permanently bonded to the paper. Without this step, the toner particles are only held in place on the paper by electrostatic and dispersion forces and are easily smeared by the slightest contact. The principle of fusing is to make the toner particles meld with each other and, by penetration into its constituent fibers, with the paper.[140,145] The toner may be softened or melted either by heat, pressure, a combination of both, or by solvent. Since toners are made from thermoplastic materials, then heat and/or pressure can soften them. Clearly this should occur with the minimum amount of energy for both economic reasons but also to avoid damage to the paper. The amount of energy needed to soften a thermoplastic is related to its rheological properties and particularly its glass transition temperature T_g. This is the temperature at which the polymer chains comprising the thermoplastic acquire sufficient thermal molecular motion that it begins to soften.[146,147] This temperature is also marked by a rapid decrease in viscosity as the temperatures rises further. Toners are generally designed to have glass transition temperatures lying between 50°C and 65°C. The reason for this rather narrow window is that if T_g is less than 50°C then toner clumping or caking is likely to occur under anticipated ambient conditions. On the other hand, if T_g is too high then excessive amounts of fusing energy will be required. The glass transition temperature is a function of the molecular weight of the polymer while the viscosity of the melted toner is affected by the nature of the polymer or blends, the molecular weight and the degree of crosslinking within the polymer.[146,147] Clearly, careful considerations have to be given to these thermal and viscoelastic properties of toner materials in addition to the charging properties and the ultimate ability to fabricate particles of the correct size and distribution; a daunting challenge indeed. Representative glass transition temperatures for toners are about 60°C and fusing temperatures are about 130°C. Since the boiling point of water (100°C) is less than the fusing temperature of toner, any water absorbed in the paper must be vaporized before the temperature can reach the fusing temperature. Since the heat capacity and latent heat of evaporation of water are so high, under a normal ambient humidity of 40 percent, a significant fraction of the heat required to fuse a xerographic image is expended in removing this moisture.[51]

The earliest fusing systems, used for example in the Xerox 914, were of the radiant heat type, in which the heat was produced by incandescent lamps. Such lamps emit radiation in the 1 to 4 micron wavelength region of the infrared part of the spectrum. Fortunately since the toner is black it is an extremely good absorber of these wavelengths, whereas paper is relatively poor in this respect.[140] To further illustrate the complications of moisture, water has absorption in the infrared and, if present in considerable amounts, can compete with the toner particles for this energy. Despite the fact that radiant fusing had the attractive features of immediate, instant-on warmup and indirect contact with the toner image, these were offset by the small but real potential for a fire if a paper jam occurred in the fuser system. Another disadvantage was that great care was needed to

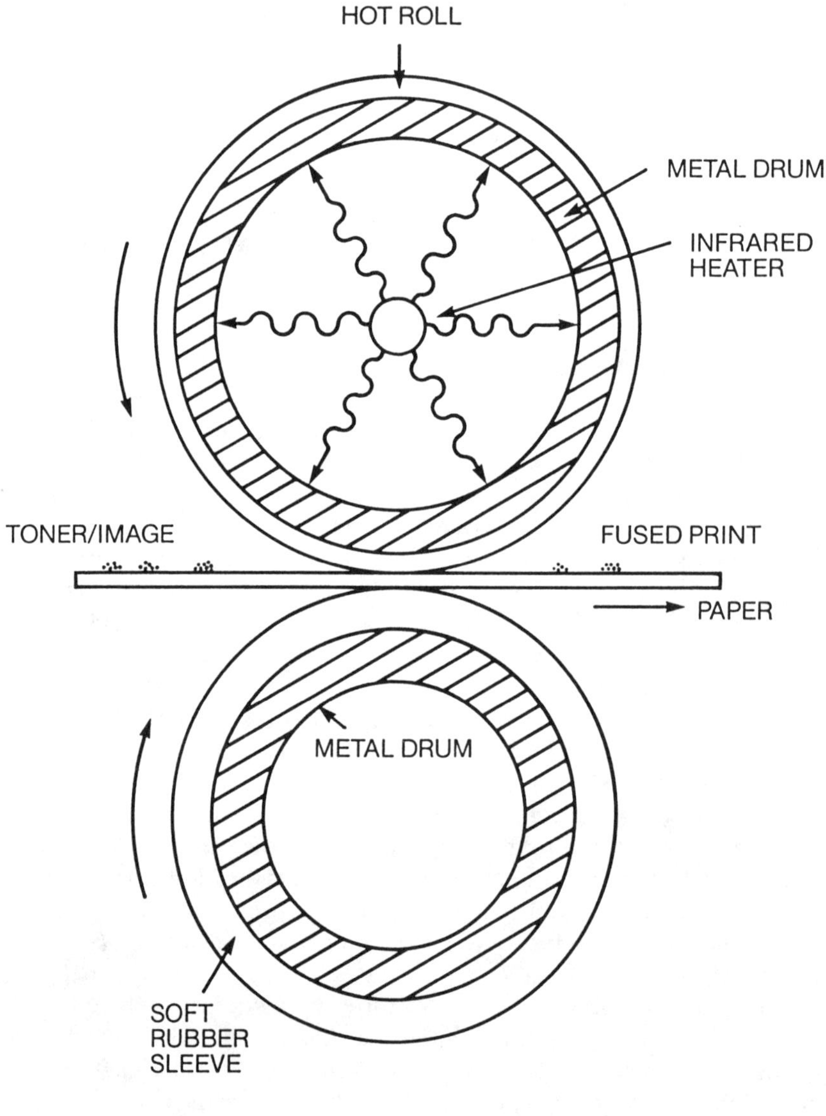

The transferred toner must be fixed or bonded to the paper. A widely used method is that of hot-roll fusing which avoids the complications of potential fires associated with radiant fusing, effluent with vapor fusing or calendering associated with cold pressure-roll fixing. The hot-roll is heated from within by some form of radiant heating and its surface is typically coated with an elastomer treated with a release agent to ensure that the molten toner does not adhere to the fuser roll.

shield the photoreceptor from stray light from the radiant fuser, particularly if the photoreceptor were sensitive to red light, otherwise light fatigue could occur. Almost all current copier products use hot-roll fusing in which a combination of pressure, shear force and temperature are used. Hot-roll fusing operates at a lower temperature, there is no associated visible light, it can handle a wider range of paper stock and can fuse effectively and more efficiently at high speeds. One problem peculiar to this fusing system is that of offset. This is caused by melted toner building up on the hot roller, which is typically a metal drum coated with a silicone rubber. The offset is ameliorated by using a release agent such as a silicone oil on the roller.[148]

An alternative fixing process is cold-pressure fixing. This process involves no direct heating so there is no warm-up time and the energy requirements are obviously lower. Hardened steel rollers which have a high quality surface finish and very low run out must be used. Nip pressures between the rollers must be at least 100 pounds per linear inch. The temperature rise associated with this is not likely to lead to significant softening of the toner so that the fixing is mostly a consequence of cold-flow. This means that softer toners (e.g., polyethylene type polymers) than those used in hot-roll fusing must be employed. Unfortunately the enormous pressures involved cause permanent calendering of the paper, in which the thickness of the paper is actually reduced as air is excluded and the fibers compressed and the toner image often characteristically acquires a gloss.[140] Other associated complications can include smudging of the image, toner adhesion to the rollers and the need for paper stripping.

Although toner transfer is a highly efficient process, approaching 90 percent there is residual toner perhaps amounting to several milligrams left behind on the photoreceptor surface. In preparation for the next cycle, this must be removed.[51] Without this cleaning step, the toner remaining affects the succeeding corona charging and image exposure and does so in a progressively deleterious fashion, as more toner is accumulated in succeeding cycles. First of all the toner acts as seeding sites for the entrapment of further toner and, because of its opaqueness to visible light, prevents the exposure light from striking the photoreceptor. Conceptually, any number of ways can be conceived of to perform this cleaning function but basically two approaches have evolved into common usage.[50] Both involve direct physical contact with the toner and photoreceptor, as opposed, for example, to the use of an airjet. The first involves the use of brushes while the second uses a blade as the cleaning element. As in the case of transfer, the residual toner is bound to the photoreceptor surface by a combination of dispersive and electrostatic forces and the physical force exerted by the brush or blade must overcome these.[133,134,149] As we have seen dispersion forces increase with the area of contact so that careful design of the

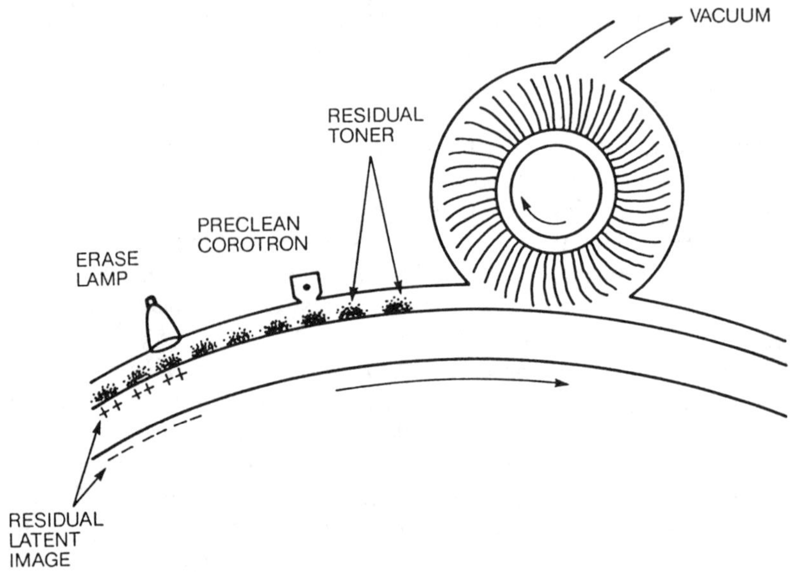

Although the transfer step is surprisingly efficient, residual toner together with paper and other particles can accumulate particularly over time. A cleaning step is therefore provided. A variety of techniques have been employed singly or in combination. These include brushes of various materials, vacuum systems and blade cleaning.

preceding steps of development and transfer to avoid the deformation of toner, is important in minimizing the magnitude of the cleaning problem.

In the case of brush cleaning, many design factors have to be considered such as the elastic properties of the synthetic bristles and their density; in the early days of xerography, rabbit's fur was found to work exceptionally well.[150] The velocity of brush rotation is also important and speeds over 1000 rpm are not uncommon. With blade-cleaning, similar care in the choice of materials is critical but there are additional factors to take into account. For unlike brush-cleaning, the sharp blade must conform closely and touch the surface of the photoreceptor.[151] The design engineer must consider that toner particles as small as one micron or less must be removed requiring the maintenance of close tolerances over the full length of the photoreceptor. For this reason conformable elastomers are typically used, although with harder photoreceptors like amorphous silicon, thin steel blades can be employed. In either case, the angle of attack of the blade and the frictional forces between it and the photoreceptor must be carefully considered. Anyone who has experienced balky windshield wipers in an automobile with the resultant chattering and streaking can appreciate the nature of the problems of blade-cleaning photoreceptors. As before the utility and efficiency of blade-cleaning is itself affected by the properties not only of the toner but also the photoreceptor. In particular blade-cleaning puts a premium on the smoothness with which the photoreceptor can be fabricated since any asperities on the surface will lift the blade locally and allow toner to pass under and escape removal. Lubricating additives such as zinc stearate are used to reduce the adhesive forces between toner and photoreceptor and allow the blade and brush to function more easily and smoothly.[142] It is also possible to use preclean corona charging or exposure to reduce the electrostatic binding forces.[152] In both cleaning techniques, the removed toner must be further transported away from the brush or blade to avoid redeposition or the impeding of further cleaning. This is typically done by using what in essence is a vacuum cleaner.[150] The cleaning step involving as it does direct physical contact places an additional burden on the photoreceptor so that in addition to the demanding electrical and photoelectronic properties, there is now the requirement that these properties and its mechanical and chemical stability withstand the considerable stresses exerted by the cleaning subsystem.

At this stage the original latent electrostatic image produced by the projection of an optical image on the precharged photoreceptor has been developed by toner, the toner transferred and residual toner cleaned. Still remaining, however, is some remnant of the original latent electrostatic image. If this is not to interfere with the next (and typically different) image, then this must be removed. Failure to do this, or do it adequately, results in the well-named phenomenon of "ghosting" in which a previous image remains and subsequently prints out. Removal of this latent image is usually achieved by uniformly exposing the photoreceptor to light from an erase lamp. This uniformly photodischarges the entire surface of the photoreceptor. The choice of erase lamp intensities and wavelengths have to be personalized to the particular characteristics of the photoreceptor in use. The intensity required may be reduced by using a preerase corotron which deposits oppositely-charged corona ions to those used in the original charging step.

The development process used in xerography is remarkable in many respects. It utilizes one of the oldest manifestations of electrical charge, triboelectricity, which, with hindsight, was one of nature's ways of revealing the structure of the atom. Despite this antiquity and the modern tools of science it is also a phenomenon for which a detailed microscopic explanation is still lacking. The two-component development process involves the interplay of triboelectricity and adhesive forces both of which are highly surface sensitive with regard to impurities and ambient humidity. Typical toner particles carry a surface charge density of 20 nanocoulombs per square centimeter which means that only about one in 100,000 of the surface atoms of the toner material are involved. Given these facts the degree of reproducibility that can be achieved in the manufacture of toner and its demonstrated reliability of operation in the market place is amazing.

Chapter 5
Laser Printing: A Light Pen

There is a basic motivation in man to record his ideas, thoughts and emotions across distance and time. So strong is this urge that even prehistoric humans, dating from about 20,000 B.C., left their mark in cave paintings which pictorially recreated everyday visual experiences. By about 4,000 B.C. in Mesopotamia, the use of much smaller images or pictographs, later called hieroglyphics by the ancient Egyptians, were used to represent individual objects. Communication eventually began to use modified pictographs for the expression of abstract concepts such as time and emotions. The final step was the establishment of an alphabet which is the use of sound or phonetic symbols. The precursor of our modern day alphabet had its origins, first with the Phoenicians around 1200 B.C., and then the Greeks. In fact the word "alphabet" comes from "alpha" and "beta" the first two letters of the Greek alphabet. These in turn were derived from the Phoenician "aleph" and "beth" which meant "ox" and "house," respectively; if the capital letter "A" is inverted, it is not hard to see an ox.[153] It was probably no accident that Phoenicia, the mercantile center of the world, developed an alphabet as a much more efficient way of recording and therefore conducting business. By contrast the Greeks, who actively traded with the Phoenicians, saw the 19-letter alphabet principally as a tool for preserving the knowledge that they sought and prized so highly. Over a period of several hundred years, the Greeks introduced additional letters, including vowels and, by dropping several of the Phoenician consonants, arrived at a 24-letter alphabet. The fact that today, in the West, writing and reading goes from left to right was also instituted by the Greeks, although their writing only used capitals with no space between words or punctuation. The Etruscans, who had adopted the Greek alphabet, were the first rulers of Rome, so that it was the Romans who essentially gave us our modern alphabet; they used "V" and "I" as both vowels and consonants and the letter "J," derived from "I," only appeared on the scene in the fifteenth century. Thus over the course of only a few thousand years communication of the complexities and subtleties of human experience by written words became possible.

Although this development was of the highest importance in the evolution of mankind, the situation was not without its initial drawbacks.[154] In the first place the number of people who could read or write was minuscule; nor was there a widespread availability of documents, let alone books. In fact the major producers and users of all written information were governments and the church. As Christianity spread the church became the guardian of education and the source and repository of writings. The sacred texts were copied by monks who often spent their whole life at the task of printing with pen and ink. Thus a major occupation of the Middle Ages was that of a scrivener whose task was to manually reproduce or copy manuscripts. These scriveners, based both on the drudgery and artistic challenge of their everyday work, were literally to determine the shape and form of the letters to be used. Thus simply to increase the productivity, the original use of only capital letters gave way, except for the start of a sentence, to the smaller and simpler lowercase letters. On the other hand, the style of the letters or fonts was very much influenced by the tools used and raised to a high artistic standard. The shape of classical Roman letters originated in the wide, thin brushes that were used to form the letters on stone before chiseling began, while the broad vertical and narrow horizontal lines and the use of serifs to cap the vertical strokes were initially used to simplify the stonemason's task.

For about 1500 years after the development of alphabets, there were no significant advances in the means of communication, apart from the combining of liquid inks and writing instruments with papyrus and parchment. The fact that both paper and printing were known in the Orient as early as the first century makes it curious that the explosion of printing had to await the diffusion of these technologies to Europe. Thus by 950 A.D. money was being printed on paper in China by the use of wood blocks. The lack of printing of the written word in the Orient is attributed to the fact that the Chinese and Korean alphabets were based on ideas or ideograms for which there were literally thousands of symbols. The spark therefore was the marriage of the wood block printing with the much more flexible and efficient phonetic Latin alphabet. This, together with a growing pressure to communicate more effectively, widely, and faster, led to the search for increasing ways to copy or reproduce documents more rapidly and at an affordable price.[10,154]

This was the situation at the end of the fourteenth century when Johann Gensfleisch zum Gutenberg was born in Mainz. It is often said that Gutenberg printed the first book, the 42 line or Gutenberg Bible, as distinct from manuscripts, but the format of books as we know them today date to the first century A.D. with the introduction of the codex. This was a collection of parchments bound on one edge and was typically produced in monasteries. Gutenberg's real contribution centered on the production of a movable type and although there is debate as to whether the invention was due to others, this uncertainty merely underlines the intensity of the search for printing advances at the time. Gutenberg was a goldsmith and it was these skills that naturally influenced him to produce movable type by cutting it, in reverse, in the form of a punch which was then used to produce a metal mold. Not unnaturally he chose to use the most popular manuscript lettering in his day and country which was northern gothic. This had evolved from the use of quill pens which, unlike brushes, lacked the flexibility to easily reproduce roman lettering. This is but one example of the evolution of new letter types based on the peculiarities of the writing instrument. Having produced the molds, the individual letters were produced by filling the mold with a low melt, tin-bismuth alloy so that Gutenberg had individual letters from which he could compose a page or a book; it is reported that it took five years to compose the Bible. Gutenberg's printing method was a form of what today is known as letterpress printing which uses a raised image that is inked and pressed onto paper. Anyone who has used a stamp and ink-pad has practiced letterpress printing. The Gutenberg Bible appeared in 1455 and was followed by an explosion in printing and the spread of knowledge, although this did not take place without resistance. This arose from entrenched interest groups such as the scriveners, whose livelihood was jeopardized, or censorship from those threatened by the specter of the uncontrolled spread of knowledge to the masses. Still, by 1500 approximately 40,000 recorded books had been produced and movable-type printing spread rapidly throughout Europe. William Caxton transferred the technology from Europe, where he had worked, to England about 1475 and was responsible for the first printing of writings with broad popular appeal such as *The Canterbury Tales* and *Aesop's Fables.* The cultural renaissance and the political and religious reformation greatly accelerated both the publication and reading of books, since the concomitant growth in literacy led to still further demand. The discovery of the New World and its subsequent colonization was in part stimulated by the dissemination of publications and blatant promotional leaflets. As early as 1640 the Puritans established a printing press in Cambridge, Masschussetts, and by 1700 there were at least a dozen printshops in Boston. Although no more than a coincidence, it is intriguing that Benjamin Franklin, in addition to his contributions to electrostatics, also trained on a press during a two year stay in England and indeed identified himself as a printer in his last will and testament.[154]

For the next two hundred and fifty years, little progress was made and printing remained slow, costly and manual. It is estimated that the time to produce a book only decreased by a factor of two in this period. However in 1796 a Bavarian, Alois Senefelder discovered lithography.[154] Senefelder who had written plays, but had failed to get them published, decided to

The Stromberg-Carlson SC-5000 computer output printer, publicly demonstrated in 1957, imaged the display from a CRT onto a XeroX Copyflo continuous printer which was originally developed to make continuous copies from roll microfilm. The SC-5000, advertised as having an output printing speed of 84,000 words per minute, was clearly a product ahead of its time. (Photograph, Courtesy Xerox Corporation)

print them himself. As a result he began a search for a simpler and cheaper method of printing but, not unlike Carlson 140 years later, endured years of trial and error, advances and setbacks, until he achieved his goal. After a series of unsuccessful attempts to improve existing printing methods using copper, he turned his attention to Kelheim limestone found in Bavaria. By chance he discovered that water would not wet letters written with a grease crayon on a highly polished piece of this limestone, while wetting the rest. In turn an oil-based ink was repelled by the wet areas of the stone but adhered to the grease letters. The name for this technology, lithography, derives from the Greek "lithos" meaning stone. Today lithography is the most widely used printing method.[144,153] Instead of using stone, however, metal sheets are coated with chemical layers which by selective exposure to light have the same basic features of ink-repelling and attracting areas. This allows sheets to be used that are sufficiently thin to wrap around the cylinders of rotary presses and thus automate the printing process. A final step taken around the end of the nineteenth century was the use of the "offset" principle. Instead of printing directly from the metal master onto paper, the image is first offset onto an intermediate sheet, now commonly a polymer, and then transferred to paper. These advances gave photo-offset lithography substantial advantages over the older letterpress technique both in terms of speed and quality.

During the fifties and sixties as xerography was experiencing its explosive growth, research was being conducted in quite unrelated areas that would eventually radically change xerographic technology. This research was to result in the development of a pen consisting of a light beam, the laser, with which it would be possible to create documents by directly writing on the photoreceptor. Paralleling this was the systemic growth in the development and application of computers and the concomitant electronic information they produced. Given the speed at which such information spewed out, the principle way of producing hard copies was high speed, low quality and very noisy mechanical impact printers. The notion of adapting xerography for this application suggested itself in the earliest days of xerography.

As a result systems were devised to copy the images of computer-generated alphanumerics created on CRTs by a scanning electron beam. The optical image on the CRT screen was projected onto and used to photodischarge the photoreceptor in a conventional xerographic copier. Such technology was the basis of a high speed computer printer introduced in the late fifties by the Stromberg-Carlson division of General Dynamics Corporation.[155] Coded computer data were printed at rates approaching one million characters per minute using a modified Copyflo product supplied by the Xerox Corporation. While this product clearly demonstrated the basic viability of the technology, it was not a cost effective approach for any but the most highly specialized applications. Consequently it was not a commercial success, being premature in terms of need and too expensive.

The laser, which was to provide the means of avoiding the necessity to create an intermediate optical image, has its own characteristic requirements. The first is that the position of the laser beam has to be moved or scanned across the dimensions of a page and the second is that its intensity be varied or modulated in a manner related to the image to be created. The modulation can be analog, although the more preferred mode is digital since this is precisely the currency with which information processing in electronic computers and workstations is conducted. Thus, the three necessary elements, separately discovered and developed, were to meet in the 1970s: the computer to provide digital electrical signals containing the image information; the laser whose output intensity could be controlled either directly or indirectly by the computer and xerography to covert optical information into prints on plain paper.

The first laser was made by Maiman in 1960, building on theoretical predictions first made by Einstein in 1917.[156–159] If an atom is excited either by a discharge or absorption of a light photon, then an electron gains energy and moves to a higher energy level within the atom. The available energy levels are quantized so that only certain energy levels are allowed and only those photon energies which can induce such transitions are absorbed. An electron in such an excited state eventually returns to its unexcited or ground state level. The energy lost in this transition is emitted by the atom as a photon; which may or may not be visible depending on the wavelength of the emitted photon. This type of process is called spontaneous emission, since the transition occurs naturally.[24]

Einstein established the essential elements of a laser when he showed that the emission from a higher to lower energy level in an excited atom could be stimulated to occur before it would do so naturally.[159] This could happen if the excited atom were struck by another photon equal in energy to the difference in energy between the excited and ground state. Thus from one atom, two photons of the same energy would emerge and these in turn could strike two further excited atoms producing four identical photons and so on in a photon-chain reaction. This phenomenon is called stimulated emission. Einstein also showed that these photons would travel in the same direction and in phase to form a coherent beam of light. By contrast, the light from a conventional household light bulb is incoherent since its light is emitted in all directions and is out of phase; in the wave picture of light, this means that the crests and troughs of the different waves bear no relationship to one another.

It was not until 1954, however, that this effect was experimentally verified by Gordon, Zeiger and Townes who used the ideas to produce the MASER (Microwave Amplification by the Stimulated Emission of Radiation).[160] The objective was to amplify and detect extremely weak microwave radiation in radio astronomical observations. To do this they used molecules of ammonia, rather than atoms, since energy levels in molecules are closer together than in atoms and can lead to emission of microwave photons which have wavelengths of centimeters. In 1958 Schawlow and Townes predicted that it should also be practical to amplify visible light waves, whose wavelengths are a million times shorter, by a similar process using atoms.[161]

Maiman developed the first LASER (Light Amplification by the Stimulated Emission of Radiation) using a ruby.[158] Ruby is an impure form of aluminum oxide in which some of the aluminum atoms have been substituted by chromium atoms. It is the absorption of light, leading to transitions to excited states of the chromium atom, which give rubies their valued rich, red color. In particular there is an energy level 1.8 eV above

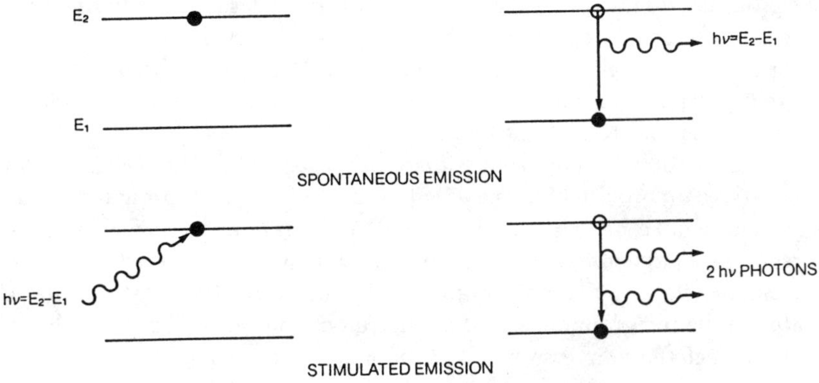

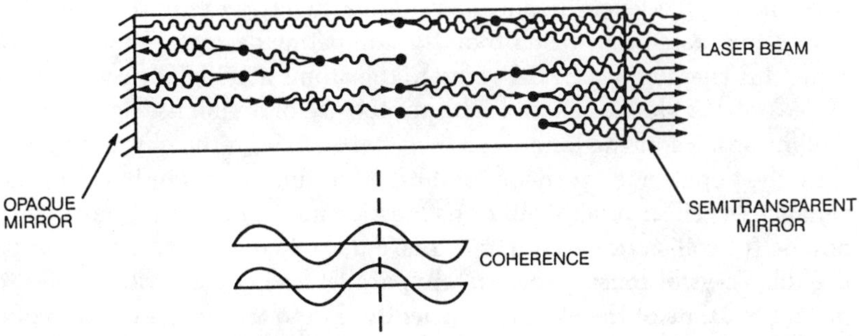

Laser which has become an accepted part of the English language is an acronym for Light Amplification by the Stimulated Emission of Radiation. Spontaneous emission of radiation is a natural consequence of thermodynamics and the structure of atoms and molecules. If an electron is initially in a level with an energy E_2 which is larger than its ground state energy E_1, it will normally drop after a very short time into the ground state, which is the highest populated state in the absence of excitation, and emit a photon of frequency v where $h\,v = E_2 - E_1$. In certain materials the time an electron stays in an excited level such as E_2 can be unusually long and the level is said to be metastable. If this can be combined by a sufficiently high excitation rate it is possible to make the level E_2 be the most populated and population inversion is said to have taken place. By excitation with a photon of energy E_2-E_1 it is possible to stimulate the electron to decay back to the ground state so that another photon of energy E_2-E_1 is emitted which travels in phase in the same direction as the stimulating photon. These two photons can induce emission of two more and so on in a kind of chain reaction so that gain can be achieved.

the ground state that plays the dominant role in the laser action of the ruby. The unusual character of this excited level is that an electron can stay in it for an exceptionally long time, thousandths of a second compared to the more typical excited state lifetime of billionths of a second. Such a long-lived state is said to be metastable.[24]

In order for stimulated emission to occur, a photon of a certain energy must strike an atom which is excited and primed to emit a photon of the same energy. To have a good chance of such event occurring, there must already be a large number of excited chromium atoms with electrons in the metastable state. Since under normal circumstances the most populated state will be the ground state, the required conditions will not usually be met. In fact the converse needs to be true, in that there must be more chromium atoms with electrons in the metastable state than the ground state. To achieve this so-called population inversion, Maiman used a flash tube to produce a short burst of intense light to pump electrons into the excited metastable level.[158] Since population inversion requires that excitation to this state occur faster than the competing decay rate to the ground state, this energy must be delivered to the atoms at a high rate and the rate of decay be as slow as possible. Hence the use of an intense light flash and the importance of the long-lived metastable state in the chromium atom. With the population inversion established, lasing can occur but in order to enhance the effect other conditions must be met. In order to maximize the number of collisions between photons and excited atoms, the two ends of the ruby crystal must be accurately parallel and coated with a reflecting metal film. One of the films is sufficiently thin to transmit a fraction of the beam and so allow it to escape. Most of the light, however, is reflected to and fro between the mirrored faces. This effectively increases the distance traveled by the light, the number of collsions, and therefore the intensity of the laser output. The color of the ruby laser is red because the energy difference of 1.8 eV between the metastable and groundstate is corresponds to wavelength of 6943 Å .[155–158]

It is beyond the scope of this volume to pursue the development of laser technology further, although the basic concepts just discussed are generally applicable to the various laser technologies now in existence. In addition to the pulsed laser, of which the ruby laser is one example, continuous lasers utilizing gases, liquids and solids are now widely available. Their output wavelengths cover the spectrum from the ultraviolet through the visible into the infrared.[156–157] The workhorse of lasers is the helium-neon gas laser which is widely used, for example, in reading bar codes at supermarket checkouts.[162] More recently solid-state lasers made from gallium arsenide which are very small and cheap have become commercially available and for laser writing have other advantages.[163] Both solid-state and gas lasers, such as helium-neon, do not need the additional light

For comparison a helium-neon gas laser is shown with the much smaller solid-state diode laser to its left; both deliver a light output of about 3 mW. The helium-neon laser emits photons with a wavelength of 6328 Å , whereas the gallium arsenide diode emission is at 8100 Å which lies just beyond the visible spectrum in the near infrared. In laser xerography, therefore, photoreceptors with sufficient photoconductivity at the relevant wavelengths are required. The gas laser must be used with a crystal modulator whereas the diode laser output can be directly modulated by controlling the current through it.

source pump for population inversion, since this can be achieved with much simplification and saving of space, weight and cost by using electric fields.

As a light pen to write on the photoreceptor a laser has many advantages. The first is the small, focused diameter of the laser spot which can be achieved, thus enabling high resolution imaging. A further consequence of the highly focused beam and its coherence is that the energy density can be very high. As a result the exposure requirements to photodischarge a xerographic photoreceptor over the dimensions of the laser beam size at very high writing speeds can be met. Secondly the light is monochromatic, that is, almost purely of one wavelength, so there are none of the complications of having the distribution of colors of conventional incandescent or fluorescent light sources. These can require expensive optical components to avoid distortions caused by wavelength-dependent properties. A third

The fact that digital printing involves writing with discrete pixels, as opposed to the continuity involved in manual writing, causes an effect called colloquially, "jaggies." Although square pixels have been used to emphasize the step-form edges produced at angles or curves, the effects are still evident with the curved pixels typically used in laser printing. As the number of pixels per inch is increased, the fidelity to the intended image improves, although still limiting on enlargement of an existing print. A variety of ways can be employed to reduce "jaggies" such as overlapping pixels or the use of gray tone densities intermediate between the white and black of binary writing. This can be done either by writing gray pixels using controlled exposure to produce partial discharge of the photoreceptor, or by electronically half-toning the digital data prior to using the conventional binary black or white of xerographic printing.

virtue is the collimated nature or absence of divergence with a laser beam; on illuminating the moon, for example, a laser beam size may only grow to about 3 feet after traveling 250,000 miles.[156,157] Thus the laser beam has all the attributes of an excellent light pen in that it is very sharp, has high intensity and is both monochromatic and highly collimated. For laser xerography the choice of laser depends first on its spectral output compared with the spectral sensitivity of the intended photoreceptor, while its power is determined by the intended writing speed. Initially the discussion will assume the use of a helium-neon gas laser which is commonly used and has an output wavelength of 6328 Å. Later solid-state lasers will be discussed since they permit considerable simplification in the laser writing system and reduction in its cost.

The laser printer has much in common with the operation of a television set in so far as both involve scanning a surface line by line.[164] By contrast to television where an electron beam continuously scans across the inside of a phosphor-coated cathode ray tube, in a laser printer a laser beam is scanned across and writes directly, but discontinuously in a series of light pulses, on the surface of the photoreceptor. The resolution capability and quality of the created document, consequently, depend on the size of the light spot striking the photoreceptor and of the total scanning system. Together with the process speed, these are determining factors in the design of the system. A letter-quality image is produced if the system can create 300 resolvable spots or picture elements (abbreviated to pixels) per inch in both the scan and process speed directions. Lower resolutions, say 120×120 pixels per inch, give lower quality prints which are, however, perfectly acceptable for many applications. The lower quality is associated with the visibly detectable jagged edge structure associated with the size and shape of the individual pixel marks. This image defect is commonly referred to as "jaggies."[165] Other factors such as alignment of characters can also mar the perceived quality but, all other factors being equal, resolution is the determining feature.

With a standard, letter-size, sheet of paper, 8.5×11 inches, and a scanner producing a pixel density of 300 per inch in both directions, each page contains about 8 million pixels.[51,164,166] If the printer is to produce prints at a high rate of 2 pages per second, the scanning system and solid-state electronics must supply the electronic information at frequencies of 16 million pixels per second to appropriately modulate the laser intensity. In fact, other requirements such as retrace of the laser beam and skipping of interdocument gaps require an additional increase of this operating frequency or bandwidth. Assuming this to be 20 million pixels per second, the laser dwell at any pixel or point on the surface of the photoreceptor is only 50 thousandths of a millionth of a second. During this fleeting time, the number of photons from the laser absorbed by the photoreceptor must photodischarge it sufficiently to create the development fields necessary to produce a dense image. Typical photoreceptor exposures to meet these conditions are about 10 ergs per cm^2.[74] Thus writing an 8.5 inch width sheet of paper at a process speed of 2 pages per second, or 20 inches per second, or 6000 scans per second would need $11 \times 20 \times (2.54)^2 \times 10$ which approximately equals 14,000 ergs per second or about 1 milliwatt of laser power. Various losses within the scanning system reduce the delivered power at the photoreceptor surface to only about 10 percent of the actual laser power, so a 10 to 15 milliwatt laser would be required. A He-Ne laser typically has an efficiency for conversion of electrical input power (watts consumed) to light output power (watts produced) of about 0.05 percent. Thus to produce 10 milliwatts of 6328 Å laser light takes about 20 watts which is what the power company charges for. Nonetheless this requirement is quite modest considering the light intensity derived from an ordinary 20 watt household light bulb. This discussion refers to a laser printer producing 120 prints per minute so that the laser power requirements would be smaller still for printers operating at more modest speeds. Likewise the operating frequency or required data rate would drop to about one million pixels per second to produce about 10 prints per minute.

Since photons are uncharged, they cannot be scanned using electrical and magnetic fields. Instead, the most commonly used technique is to use a rotating polygon mirror by directing the laser spot onto the reflective surfaces of a multi-faceted polygon which is rotated at high speed.[166] Seen from the photoreceptor's viewpoint, this looks like the output from a very high speed lighthouse. The use of some very simple arithmetic gives some idea of what is involved. The number of resolvable spots, N_r, that can be produced by a polygon with, f, facets each of width, W, for light of wavelength, λ, is $N_r = 1.26\ W/F\ \lambda$. For W=6.67 millimeters, $\lambda = 6328$ Å (He-Ne wavelength) and F=36 facets, then N_r is about 3600 pixels. In order to achieve 6,000 scans per second (each facet produces one scan line) then the polygon must rotate at a little over 166 times per second or ten thousand revolutions per minute. While certainly fast, this is now routinely used in current high-speed xerographic printers. For the polygon under consideration, 36 facets each 6.67 millimeters wide means a polygon of overall diameter of about three inches. It is almost obvious that precision engineering is of critical importance in achieving the necessary smoothness of rotation and reflection uniformity of the laser light from the facets of the polygon. The velocity of the laser beam at the surface of the photoreceptor is about 1.5×10^5 cm/s or 3,000 mph. Even in a high speed machine, where the photoreceptor rotates at right angles to the scan at a process speed of only about 65 cm/s (1 mph), the photoreceptor is thus effectively stationary during any particular laser scan.

Having described the means by which the laser spot is scanned across

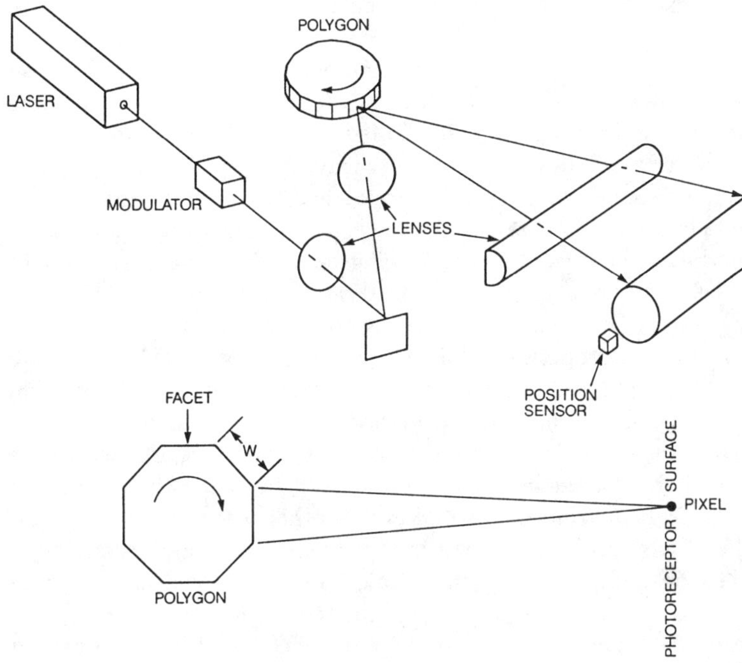

Electrons in a television set are scanned across the CRT by suitably modulated electrical and magnetic fields. Since photons are uncharged, writing on a photoreceptor in a laser printer is accomplished by scanning the laser beam using a rapidly rotating set of reflective mirrors in the form of a polygon. The number of mirrors or facets in the polygon and its rotational speed determine the resolution or pixels per inch that can be addressed and the time to write a page or the scan rate. High-speed photosensors on either end of the photoreceptor provide signals as to the position of the laser beam to within a fraction of a pixel.

the dimensions of a page, attention must now turn as to the manner in which the information is impressed on the laser beam. This is accomplished by arranging for all or none of the laser light to reach the photoreceptor. Every spot or pixel is therefore either exposed or unexposed and is subsequently developed as black or white which has a certain natural compatibility with the binary aspect of computer technology. The discussion of gray level printing for pictorial output is deferred for convenience to the next chapter. The image data in the form of digital voltage pulses may originate from any number of sources including work stations or word processors, magnetic memory files or a scanner, and these in turn may be dispersed over a linked network. The digital or video signals cannot be used to directly modulate the laser since it is impossible to turn to gas laser on and off at a rates of millions of times a second. Instead the signals are amplified and applied to a device called an acousto-optic modulator cell.[51,167] In such a cell, ultrasonic pressure waves are created by using sonic transducers or sound-wave generators attached to a crystal such as lead molybdate. In this and other materials, such propagating sound or pressure waves modulate the crystal lattice spacings and can interact with a light beam traveling at right angles to the direction of propagation of the ultrasonic beam. These variations arising from pressure-induced fluctuations in the optical properties of the lead molybdate effectively constitute a diffraction grating.[167] In the absence of any sound waves, the laser beam propagates with little or no interaction through the transparent crystal (on signal). By transmitting a sufficiently intense sound wave, related to the magnitude of the voltage associated with the digital data, the light beam is diffracted or bent from its normal path so that, for example, no light strikes the photoreceptor (off signal). The intensity of the diffracted light is related to the amplitude of the ultrasonic wave while the degree of bending is proportional to its frequency. Another consideration is the speed of response of the acousto-optical modulator, since this limits the rate at which the laser can be modulated and data handled. With careful design, acousto-optical modulators are capable of handling up to 100 million pieces of data per second which is consistent with the bandwidths required by the current pixel resolutions and process speeds.[166]

The discovery in 1962 of the solid-state laser based on gallium arsenide semiconductors was an important step forward, although the initial devices were limited to operation in the infrared and at very low temperatures (−196°C).[163] During the sixties and seventies, considerable progress was made leading to the development of solid-state lasers that operate at room temperature with long life and high output intensities at wavelengths moving into visible wavelengths. From the perspective of laser printing, the main advantage these devices offer is that the light intensity output is directly controlled by the current passed through the semiconductor p-n junction. This introduces a major simplification, since an acousto-optical modulator is no longer needed. To date, however, the shortest commercially available laser wavelength, apart from experimental laboratory prototypes, is at 7800 Å, so photoreceptors insensitive to this wavelength such as amorphous selenium cannot be used. Fortunately organic photosensitizing layers, used in dual-layer photoreceptors, or amorphous silicon have useful photosensitivity at 7800 Å.[106,118]

Another approach to constructing a xerographic printer is to use a high-density linear array of LEDs, Light Emitting Diodes.[168,169] These are

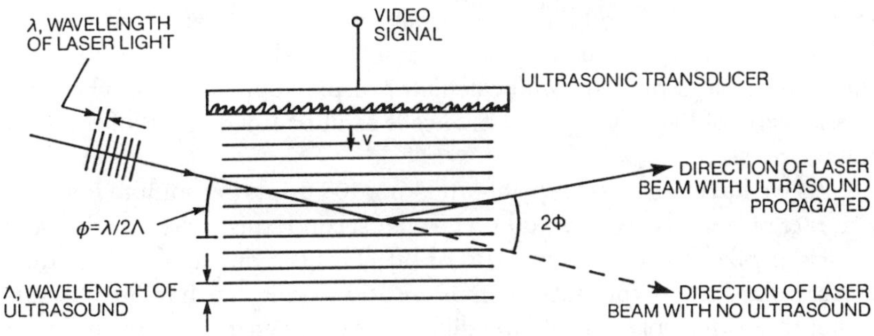

Gas lasers cannot be turned on and off very fast so that the video signal, which calls for particular pixel sites on the photoreceptor to be exposed or not to the laser beam, is applied to a device known as an optical modulator. This consists of crystals whose optical diffraction properties are changed by the application of ultrasonic waves of wavelength Λ *induced in a transducer by the video signal. In this way the laser beam can be bent through an angle 2* ϕ *so that it only illuminates the photoreceptor when required.*

also gallium arsenide-based solid-state devices which, although not lasers, can be produced in high-density arrays up to a page width. To achieve a full page array of 300 pixels per inch means that the linear array must contain several thousand individual LEDs and high efficiency self-focusing lenses are required to deliver as much of the emitted light as possible to the surface of the photoreceptor. Page-width LED image bars obviously eliminate the need for both a rotating polygon scanner and a modulator.[168] The simplification afforded is offset by a number of potential trade-offs. First the viability of the whole array hinges on the lifetime of individual diodes. Thus if only a few of the diodes fail the whole array may need replacement. Further the several thousand individual light elements must be addressed, although the use of multiplexing schemes can reduce the number of direct hard-wired connections to a manageable level. Moreover although compatible with black and white alphanumberic printing, uniformity of the light emission over several thousand diodes is needed for the higher quality printing demanded by combined text and graphics. In another related approach, the light output from an extended, stationary light source can be modulated by a page-width linear array of discrete pixel-size shutters fabricated using liquid crystals.[170] These utilize the phenomena, used in watch and calculator display applications, where the transmission of light by liquid crystal cells can be turned on or off by short voltage pulses.

As we have seen a laser printer making 100 prints per minute involves a bit rate of roughly twenty million bits per second; this is the rate that any particular pixel has to be turned on or off. This is a much faster rate than any standard microcomputer communications link can handle. To surmount these kinds of transmission difficulties data generators use character codes. In other words, instead of sending a full pixel by pixel description of each character a word processor, for example, uses a short hand code thus drastically reducing the data rate and channel bandwidth necessary. At this point some definition of terminology is both necessary and helpful.[171,172] Data and information processing by computers use a so-called binary language. This is a natural use of the two states of a pulse in an electronic circuit, i.e., either there is a pulse (denoted by 1) or there is not (denoted by 0). The 0 or 1 each represents one element of binary information and is called a bit. Just as in Morse code, where different combinations of dots and dashes represent individual characters, so codes based on binary on/off signals have been developed. To represent uniquely each of the 26 letters in the English alphabet requires 4.7 bits, since $2^{4.7}=26$; since the number of bits must be a whole number, 5 bits would be necessary. However, the full set of characters of a given size and typeface required for English text, called a font, includes lower as well as upper case letters. To accomodate a font together with numbers and other symbols such as the comma, question mark and so on a 7 or 8 bit code, is used. A widely used code representing a string of alphanumeric or graphic symbols is known as ASCII (American Standard Code for Inoformation Interchange). In turn, this forms part of a page description language (PDL) containing coded instructions for the page format, the characters designating the text on the page, the type and size of font and any patterns relating to line graphics.

As in Morse code there is no information in the ASCII code as to what a letter "A" looks like or how it can be created. Unlike Morse code, where the human operator manually translates digital dots and dashes into alphanumeric text, at some point the character code has to be translated into a video signal which can be used to drive the laser printer. The coded instructions from the source first pass through interfaces which are devices to ensure that the input information from the source computer is in a language understandable by the controller unit in the printer. The controller is an independent computer within the laser printer which deciphers the coded page description language and creates the video signal. At the time of printing, this video signal drives directly the light source in the case of

7-BIT ASC II CHARACTER CODE

A	1 000 001	a	1 100 001
B	1 000 010	b	1 100 010
C	1 000 011	c	1 100 011
D	1 000 100	d	1 100 100
E	1 000 101	e	1 100 101
F	1 000 110	f	1 100 110
G	1 000 111	g	1 100 111
H	1 001 000	h	1 101 000
I	1 001 001	i	1 101 001
J	1 001 010	j	1 101 010
K	1 001 011	k	1 101 011
L	1 001 100	l	1 101 100
M	1 001 101	m	1 101 101
N	1 001 110	n	1 101 110
O	1 001 111	o	1 101 111
P	1 010 000	p	1 110 000
Q	1 010 001	r	1 110 010
R	1 010 010	s	1 110 011
S	1 010 011	t	1 110 100
T	1 010 100	u	1 110 101
U	1 010 101	v	1 110 110
V	1 010 110	w	1 110 111
W	1 010 111	x	1 111 000
X	1 011 000	y	1 111 001
Y	1 011 001	z	1 111 010
Z	1 011 010		

0	0 110 000
1	0 110 001
2	0 110 010
3	0 110 011
4	0 110 100
5	0 110 101
6	0 110 110
7	0 110 111
8	0 111 000
9	0 111 001
!	0 100 001
$	0 100 100
%	0 100 101
&	0 100 110

The 7-bit ASCII code which is nothing more than a Morse code where 0s and 1s replace dots and dashes.

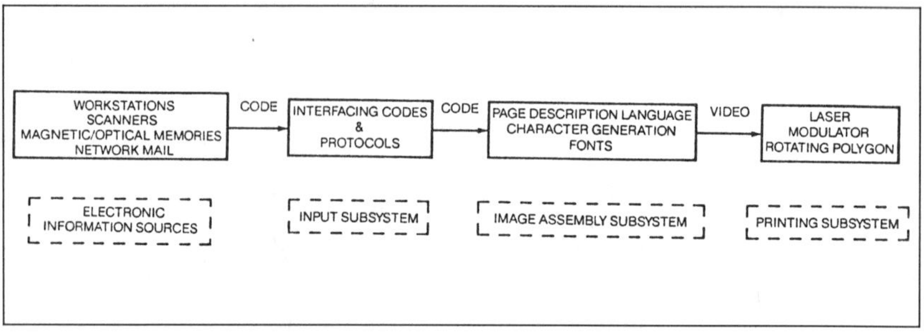

Schematic breakdown of the various subsystems involved in xerographic printing.

a solid-state laser or diodes, or the acousto-optical modulator when a gas laser is being used. The video signal contains pixel by pixel instructions which determines what the laser writes on the photoreceptor. Since the photoreceptor is an intermediate for paper at this point, it is also important that the position of the information the laser writes be positioned exactly as required in the final print. This requires precise knowledge as to the position of the laser beam in its scan relative to the release of the video data. This accomplished by using photodetectors positioned at known locations on either side of the photoreceptor, one to detect the start and the other the end of a scan.[166] Since a pixel dwell time is about 50 nanoseconds then positional knowledge within a fraction of a pixel dimension requires the use of a detector with a speed of response time of 10 nanoseconds or better. This is readily achievable with silicon photodiodes.

In laser xerography the fonts are written with light rather than carving a character in relief, casting metal from a mold, or using pen and ink. This can be accomplished in two general ways, both involving the science of digital typography which is the basis for such font design.[165] The video signal may consist of a stored bit-map description of each character which can then be created in printed form. Ideally every character should be defined as a sufficient number of pixels with a unique set of coordinates to assure the necessary resolution and hence quality. The more complete the description, however, the more demand is placed on limited and expensive memory capacity. Resolution is really only critical at the periphery of a character since, for example, the essence of a circle can be perceived even at very low resolution. By contrast at too a low resolution it may be difficult to differentiate between, for example, a "c" and an "e," or even at moderately high resolution the full details of an intricate font may be lost. This is the basis of a different type of font design, called outline or letterform fonts. This involves using either what is called run-length encoding in which only the end points of each stroke of the laser light pen are used,

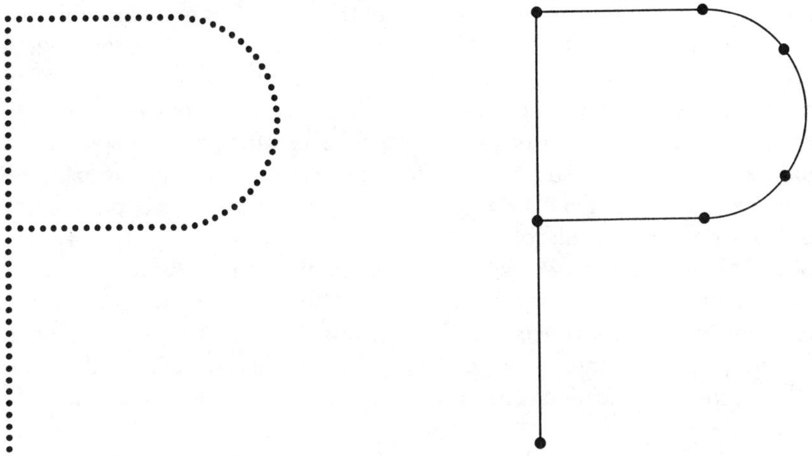

Ideally every character of a font should be defined as a sufficient number of pixels to assure the necessary resolution and hence quality. Various approaches to font design involve trade-offs between the required quality, minimizing the need for memory storage and the necessity of increased computation and hence slower processing rate. One type is outline or letterform fonts. This involves using either what is called run-length encoding, in which only the end points of each stroke of the laser light pen are used, or splines which are more descriptively termed line segments. In either case only the outline of the character is defined at a few critical points until it is to be printed, at which time the microprocessor can interpolate or connect the points with straight or curved line segments.

or splines which are more descriptively termed line segments. In either case only the outline of the character is defined at a few critical points until it is to be printed, at which time the microprocessor can interpolate or connect the points with lines. A still more complex technique is the use of algorithms or mathematical descriptions of the characters in terms of their geometric shapes. Bit-mapped fonts are typically designed for one printer and cannot be used on others which have different resolutions unless some additional processing is done to scale up or down to the new matrix array size. This difficulty is avoided with outline fonts since magnification or reduction is readily accomplished mathematically. The software design of fonts is time consuming and therefore expensive, more so for outline than bit-mapped fonts. The various approaches involve trade-offs between minimizing the need for memory storage with the necessity of increased computation and hence a slower processing rate. One ends up with a binary bit-map or raster information for each of the characters in a particular font design. These are loaded into a Programmable Read-Only Memory (PROM) which is usually part of the controller since the font bit-map can be accessed more rapidly this way.[173] The capability of the laser printer can be extended by providing the option of plug-in ROMs to complement the standard or fixed font library. If the printer has a sufficiently large Random Access Memory (RAM) buffer, it can store one entire bit-mapped page before printing commences, although usually, for reasons of cost, only a few lines of buffer storage are used. Where one laser printer is to serve as a printer for several work stations on a network then buffer storage, using a hard magnetic disk, is needed. This permits the establishment of a queue for electronic documents originating from various stations until as each job reaches the head of the line it is moved into the RAM buffer for printing.[174]

Having impressed the digital information on the light pen with modulation, the rest of the xerographic process is unchanged except in detail. Thus there is charging, exposure using the laser, development, transfer fixing and cleaning. One change can be in the mode of exposure. In xerography, originally practised by Carlson, the projected black characters from a document do not reflect light and their corresponding areas on the photoreceptor are not discharged. In laser xerography one can proceed in similar fashion in that the images to be developed are not exposed and the background areas are discharged with the laser beam. This is analogous to the way the ancient stone mason would produce letters in relief by chipping away the surrounding stone. One advantage of this approach, sometimes referred to as "positive" or charged-area development, is that it is compatible with the similarly charged toner used in a copier employing the same corotron polarities. In "positive xerography the laser can print everything including form outlines thus providing significant savings of the overhead costs associated with storing preprinted forms. Moreover the forms can be digitally stored together with the data to be inserted so that the text and form data can be combined into a single bit stream. This allows the precollation of documents and reports to be printed either on single sided sheets or duplexed in which printing is done on both sides of a single sheet. A disadvantage is that, because of the shape of the laser spot, a line written one pixel wide visually looks thinner and so at least two pixel wide lines are required. Alternatively, one can choose to write with the laser, i.e., the images to be developed black are written by the laser and the photoreceptor discharged in those areas. This is called "negative" or discharged-area xerography. In this case the toner to be deposited on the exposed or discharged photoreceptor needs a charged toner with the same polarity as the corona polarity used to charge the photoreceptor. One reason for using "negative" xerography in a printer is

The second important element in the evolution of xerographic printing was the independent invention and development of the laser and LEDs. This enabled the production of prints over a range of speed, quality and price.[164,166–169] An additional, not inconsequential, advantage of laser printing often remarked upon by users is the almost total lack of noise as compared with traditional impact printing devices, whether they be typewriters in the office or chain printers in the computer room. In conjunction with ever more affordable word processors one person can, without any outside agency, do everything necessary to take a concept from the idea stage to the final hard copy whether in the office or increasingly in the home. It is possible to print from electronically stored files which can easily and efficiently be edited, updated and if part of a network even collected from or transmitted to distant points. The magnitude of the changes underway can be gauged by the fact that today's and tomorrow's office talk is more likely to include references to file servers, pagination and icons than carbon paper or correction fluid.[176] Despite this upheaval, and contrary to some prognostications, experience has shown that the printed page or hard copy is still a favored means of communication and electronic displays, although significant, largely serve as intermediate substitutes. The consequences and implications of the continued importance of paper documents together with issues of data storage, transmission and networks will be discussed next.

Chapter 6
Digital Copiers: Read, Write and Send

In terms of document processing, we live in a hybrid world. This consists of traditional paper documents together with the increasing number that are created and exist electronically.[3,4] Whatever their origin, the vast majority of such documents result from striking some keyboard. As the previous chapter discussed, electronic information can be converted into a paper document through the aegis of a xerographic printer. The question is: can the transition be made the other way, that is, from an existing document to the electronic medium? Substantial benefits accrue from an ability to move easily in both directions. Conventional copying involves the direct optical projection of a document image onto the photoreceptor and this allows certain manipulation of the whole document image, such as reduction or enlargement, by adjustments in the optics. It has become possible to do some editing functions such as trimming, masking and centering by using coordinate grids, but these options are rather limited and not very user friendly. If, however, a document can be transposed into electrical signals then it becomes amenable, at least in principle, to two general types of electronic processing. Both involve the "reading" of the original by a transducer which produces a related electrical signal, but differ in the level of sophistication with which the document is subsequently manipulated. The first type of processing uses the electrical version of the document to transmit it to a point across the hall or around the world where an unamended paper copy or facsimile (FAX) is made on a printer. This task involves two requirements: the conversion of the document information into an electrical signal and the latter's transmission with speed and accuracy. The second and much more sophisticated electronic processing of documents permits substantial changes to be made to the electronic version of the original before the modified electronic output is sent to a xerographic printer.

Facsimile transmission or FAX, which is a rather old technology, is the transmission of text or graphics, including signatures, doodles, fingerprints, or drawings, from one place to another.[177] This form of visual image transmission easily predates its most familiar form, television. In 1843 a Scotsman Alexander Bain was awarded a British patent for an "automatic electrochemical recording telegraph."[177] Bain ingeniously conceived of the fact that two synchronized pendulums at a given point of time are in the same relative position in their motion; two garden swings moving in unison are a very good analogy. The implementation involved the mounting of electrical contacts on the bottom of each pendulum. The "send" pendulum swept over a message composed of raised metallic type and it was arranged that contact with the type completed an electrical circuit to produce an electrical pulse. Each sweep of the pendulum therefore produced a string of pulses sequentially in a time frame determined by the motion of the pendulum. On completion of each swing, the array of type was slightly moved so that eventually the total information was scanned. The contact on the "receive" pendulum swept over an electrosensitive paper which darkened whenever an electrical pulse occurred. In this way the dark marks made up a facsimile of the metal type at the "send" station. An improved version of Bain's invention was the basis of a crude FAX system linking several provincial cities with Paris in 1865.

With the observation of photoelectrical effects in the late nineteenth century, increasing attention turned at the start of the new century to the possible use of these phenomena in scanning, rather than electromechanical methods pioneered by Bain. In concept an optical image would be projected onto a photosensitive detector, literally an electronic eye, for conversion into an analog electrical signal. In fact by 1902 the first practical use of photoelectric scanning for the FAX transmission of photographs was realized.[177] A commercial system was operating between London and Berlin, as early as 1910, and in 1922 a photograph of Pope Pius XI was transmitted from Rome to the United States. In 1925 Bell Telephone scientists described a technology for transmitting FAX over existing telephone lines.[178] This was a purely telephoto process in that at the send station the image was a negative transparency in the form of a cylinder that could be rotated and moved along its axis. The negative was illuminated with a spot of light and as a result the entire image was scanned. The transmitted light entered a vacuum photoelectric cell and struck the light sensitive cathode with sufficient energy to cause ejection of electrons in numbers that depended on the light intensity. These electrons moving under the electrical field produced by a positive voltage applied to the anode electrode caused a current proportional to the light reflected from the scanned image to flow. At the receiver end the current pulses flowed through a narrow ribbon of metal covering a small light aperture in close proximity to a permanent magnet. The current pulse in the ribbon produced a magnetic field which, in the presence of the permanent magnetic field, caused the ribbon to move and allowed light to pass through onto a photographic film. This film, also in the form of a cylinder, rotating and moving in synchronization with the negative at the sender station, on development produced a copy of the original. The combining of these FAX techniques with electrostatic imaging appears to have been first demonstrated in 1938 by Selenyi.[46] The analog electrical signal produced by scanning a picture at the send station was applied to the grid of an ion source at the receive station. The receiver drum, coated with a black insulating varnish called an electroreceptor, was rotated at 900 rpm and displaced axially 0.3 mm per rotation. This gave an estimated resolutionof 80 scan lines per inch. The analog electrical signal modulated the number of ions sprayed onto the electroreceptor and produced a latent electrostatic replica of the picture which was then developed with dry powder.

Early on in the development of xerography the ability to copy at a distance over ordinary telephone lines was conceived and LDX (Long Distance Xerography) facsimile equipment, more commonly referred to today as a telecopier, was introduced in the early sixties.[155] One of the earliest approaches employed a cathode-ray tube as the source of the light spot which was used to scan the document. The reflected light from the document was collected by a light pipe and converted into an analog electrical signal by a thermionic photomultiplier tube. The analog electrical signal was then sent over commercial telephone lines where the electrical signal modultated another cathode-ray tube. Its light output was then projected onto a xerographic photoreceptor to reproduce the original document as a developable electrostatic image. In common with all FAX technology, severe limitations originally existed in quality and speed, not the least of which was the limited capability of the telephone circuits used which had been designed for voice transmission. To appreciate the nature of these difficulties and the current advances being made in techniques of data transmission this topic must be briefly discussed.

The average person can hear sound waves with frequencies between 30 to 20,000 Hz or 20 kHz. Such a range of frequencies is called a bandwidth. To transmit high fidelity sound over conventional telephone lines, called twisted-pairs, into the home, the continuous range of frequencies from 30 to about 20,000 Hz would have to be sent. Although this is technically feasible, the telephone company deliberately chooses to transmit only frequencies from 300 to 3400 Hz.[171,172] This allows the more economical use of the telephone lines so that many different messages can be sent over the system simultaneously. The restricted bandwidth and frequency response is still compatible with the recognition and intelligibility of the human voice. The transmission uses an analog system in that the electrical current in the lines varies continuously in an analogous manner to the air pressure produced by sound waves and ultimately detected by the

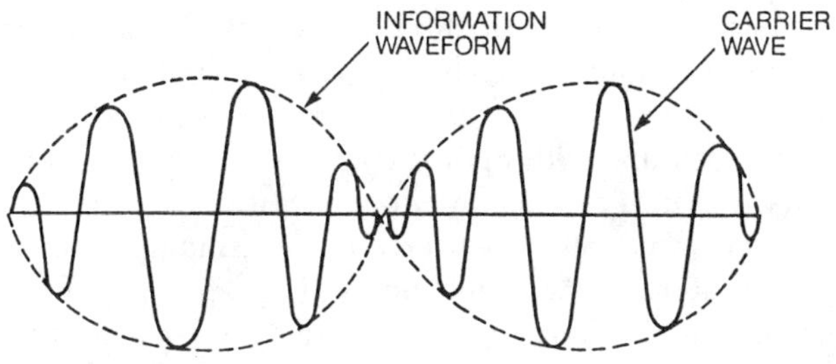

AMPLITUDE MODULATION

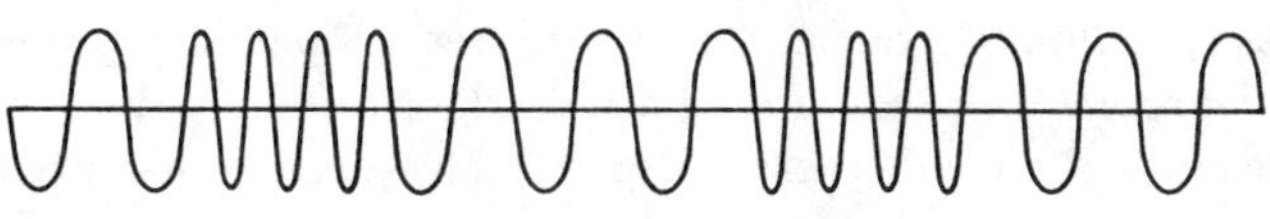

FREQUENCY MODULATION

The most common form of information transmission by modulation is used in radio. In AM or Amplitude Modulation the amplitude of a higher frequency electromagnetic wave known as the carrier wave is modulated by the lower frequency audio signal. In FM or Frequency Modulation the carrier wave frequency is modulated at the audio frequency.

ear. Since one voice channel only uses a bandwidth of about 3400 Hz, a number of channels can be transmitted simultaneously or, in other words, multiplexed. This is frequency-division multiplexing, an everyday example of which is radio broadcasting. Radios receive signals which emanate from broadcast stations at different frequencies. The audio frequencies of voices or music modulate the station frequency or carrier wave which is its position on the dial, either by amplitude for an AM station or by frequency for FM. The tuning circuits in the radio receiver demodulate the signal to extract the audio component from the modulated input. Thus AM and FM broadcasting are applications of frequency-division multiplexing in which many stations each use a different fraction of the AM and FM radio bands respectively. The number of signals that can be handled by a transmission network depends on the frequencies the propagating medium can carry. Pairs of wires can transmit, using repeater amplifiers, around 2 MHz, coaxial cables about 70 MHz and microwaves up to about 10,000 MHz or 10 GHz. Consequently, a single wire-pair typically can carry tens of voice channels a coaxial cable several thousand and microwave technology more than one hundred thousand. Light is an electromagnetic radiation with frequencies 100,000 times higher than microwaves so even greater channel capacities are possible with optical communication using fiber optics.

When thousands of telephone conversations travel either over coaxial cables or microwave links they have to be demultiplexed, switched and remultiplexed at each switching point. Repeater amplifiers, moreover, which unfortunately boost both the signal and noise must be used at relatively short intervals, otherwise the signal is attentuated and lost in the noise. As the channel capacity increases, these associated component costs become a critical economic consideration. Increasingly, data transmission using digital rather than analog signals is employed since it offers significant benefits in the quality and speed of data transmission. These advantages have a number of origins. The first is that digital pulses can be regenerated at various points along the transmission grid so that there is continual discrimination against unwanted noise. Secondly, digital transmission is much less susceptible to noise than analog since, if 0 volts represents 0 and 10 volts represent 1, it would take a noise level of ± 5 volts to introduce an uncertainty as to whether a 0 or a 1 had been transmitted. Digital languages other than binary are more susceptible to noise. Thus in a decimal or 10 level system, if each level differed by 1 volt, a noise level as low as a fraction of a volt would cause comparable uncertainty. For this reason the transmission mode of choice is binary digital consisting of on/off pulses such as those used in integrated circuits or computers. Despite these advantages and the growth in digital transmission, most of the world's telecommunication circuits are still analog in character simply because of the enormous sunk capital costs. As a result analog and digital technologies coexist. This is the reason why communication between a computer terminal and a central computing unit over conventional telephone lines requires a MODEM device. MODEM, a contraction of MODulate and DEModulate, involves the conversion of the digital data output into analog signals for transmission over the telephone line.

To encode or digitize an analog signal it must be quantized and sampled at periodic intervals by a process called pulse-code modulation (PCM)[171,172] Each sample determines the voltage of the analog signal at a given point in time. If the number of quantized levels is 128, then the sample can be coded with a 7 bit binary number respectively ($2^7=128$) and it is this stream of 7 bit binary digits that is transmitted. In the 1920s Nyquist published papers dealing with the science of information theory in which he showed that if the highest frequency a signal contains is f, then a pulse train of 2f pulses per second is sufficient to ensure transmission and faithful reconstruction of the signal, assuming no quantization error. Thus to digitize the human voice limited to 4,000 Hz requires 8,000 samples per

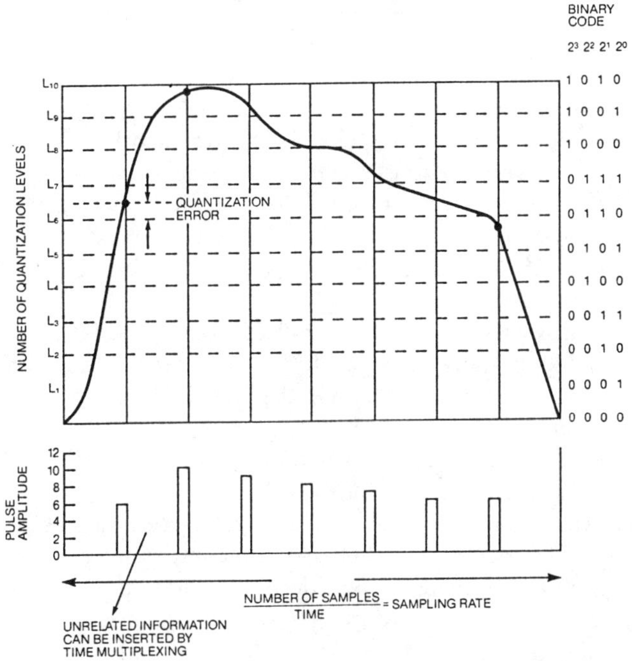

The transmission of digital information may involve PCM the Pulse-Code Modulation of an analog signal. The amplitude of the analog signal is sampled at set times and the sampled amplitude assigned a quantization value denoted by a binary number. The accuracy of the digitization depends on the number of quantization levels selected and sampling rate. In the example shown, the point is given the value L_6 *even though it lies between* L_6 *and* L_7*. Since the duration of the sampling pulse can be very short, a series of unrelated signals can be time-multiplexed by insertion and transmitted simultaneously, thus increasing the rate of data transmission.*

second or 56,000 bits per second, (bps), if 7 bits are used. Each sample digitized using 7 bits can in principle differentiate 128 volume levels. A data rate of 56,000 bps is very small compared with the transmission rates possible, since with digital repeaters about 2 million bps can be carried over a wire-pair and much more on coaxial cables and microwave circuits. In the example just considered, if the sampling rate is 8 kHz, then the pulses would occur every 125 millionths of a second. Since the pulse need only last one millionth of a second to acquire the sampling information, it is clear that more pulses from other signals can be transmitted at the same time. This time-division multiplexing, as it is known, allows the capacity and efficiency of transmission to be maximized by sending several digital signals along the same circuit. Samples from different messages mixed prior to transmission can be ultimately unscrambled and reconstructed at high speed with synchronization pulses and logic circuits. The circuit components employed in time-division multiplexing are much cheaper than those used in frequency-division multiplexing for analog transmission. One final important consideration in the transmission of information is the speed with which it can proceed. This is determined by the channel capacity expressed in bps. The capacity of a channel is, according to the well-known Hartley law, proportional to its bandwidth to faithfully transmit the information; however, higher bandwidths allow data transmission to proceed faster so that with recent developments 16 billion bps can be sent on a microwave waveguide circuit.

FAX has a significant advantage over postal or courier services in that text and graphics of more or less arbitrary form can be transmitted in matters of minutes or less to anywhere in the world, if two FAX machines can be suitably linked.[177] In contrast to computer-based electronic mail systems, FAX is much further along in terms of standardized equipment and transmission methods. As a result the different vendors in the FAX market conform to various standards established by the International Consultative Committee for Telegraph and Telephone (CCITT). Four categories of FAX machines were established though, despite the fact that several hundred thousand of the first two groups are still in operation, these are now technologically obsolete. These machines termed Groups 1 and 2 were generally analog in type and could transmit, with a 96 pixel per inch resolution, a standard-size document in about three to six minutes at 2,620 pixels per second. Group 3 machines introduced about 1980 are completely digital and since they are intended to use conventional telephone lines, referred to as Public Switched Telephone Networks or PSTNs, they must be used with a MODEM. The standard bit rate over analog lines is 9600 bits per second but depending on the quality of the telephone transmission these machines have the capability of automatically reducing the bit rate to maintain quality. Since the space in many documents is mostly white or may contain black areas of many pixels, it is possible to speed up the transmission rate by eliminating the redundancy of processing like pixels. This technique known as data compression can compress data by a 10 to 20 factor so that the transmission time for a standard one page letter becomes 30 seconds or less. The final Group are also digital machines but differ from Group 3 in that PSTNs are not used for transmission. Instead Public Data Networks or PDN's are used so that no MODEM is required and more sophisticated algorithms for data handling and compression are employed. A major advantage of PDN's is that data rate transmission of 56,000 bps or greater can be anticipated compared with the 9,600 bps rate of a PSTN. These special digital networks enable higher transmission rates and/or higher resolution and, although initially more expensive, the higher costs will trend downward. Already the combination of such networks combined with xerographic facsimile machines allow resolutions of 400 pixels/inch with a transmission time of 10 seconds or less for a typical document. Such capability produces letter quality, plain paper reproductions of any arbitrary input.

The early optical scanners used were based on vacuum-tube technology with a photosensitive cathode enclosed in a vacuum tube.[176] The scanners employed today are based on solid-state photoconductors which function on the principle that with exposure to light there is a change in their electrical properties.[36,177,179] This can either be a change in resistance (photoconductivity) or the production or a voltage across a junction (photovoltaic effect). Ideally, the changes in current or voltage should be proportional to the light intensity to ensure a faithful rendition of the density levels in the original document. Moreover, the higher the photosensitivity the better the signal to noise ratio and the faster the scanning can proceed. A scanner must also have the capability to respond to the full range of the visible spectrum, since it is expected to handle documents that contain colors in addition to black and white. The situation with respect to color sensitivity, however, involves more than the spectral sensitivity of the detector, for even if this is uniformly panchromatic, the spectral output of the illuminating source then becomes the determining factor. For example, with a laser illumination source emitting one wavelength, the scanner would be blind to most colors, while incandescent light sources provide much more light in some parts of the visible spectrum than others.

Just as printer resolutions of 300 spots per inch are required to write letter quality characters and up to 1200 spots per inch for graphic quality equivalent to offset lithography, the resolution capability of the scanner is also important. The fidelity with which the original image can be "read" depends on a sufficiently high density of individually addressable photosensitive eyes or pixels in the scanner. For the scanner must view the original document and using the photoelectrical effects, convert it into an analog electrical signal which can be digitized for electronic processing, storage or transmission to a workstation or printer. The material of choice for scanners today is crystalline silicon although, no doubt stemming from Willoughby Smith's work, one of the earliest materials considered as a solid-state photodetector was selenium. The preeminence of silicon depends on its outstanding photoconductive parameters in terms of sensitivity, speed of response and spectral sensitivity. Just as important, if not more so, is the fact that as the basis of the semiconductor industry silicon processing and manufacturing technology is fully established to produce integrated circuits of the required density and complexity at low cost and with high reliability.[179]

Currently the most widely used solid-state array scanners are Charge-Coupled Devices (CCD's).[180] A CCD functions by the use of storage locations

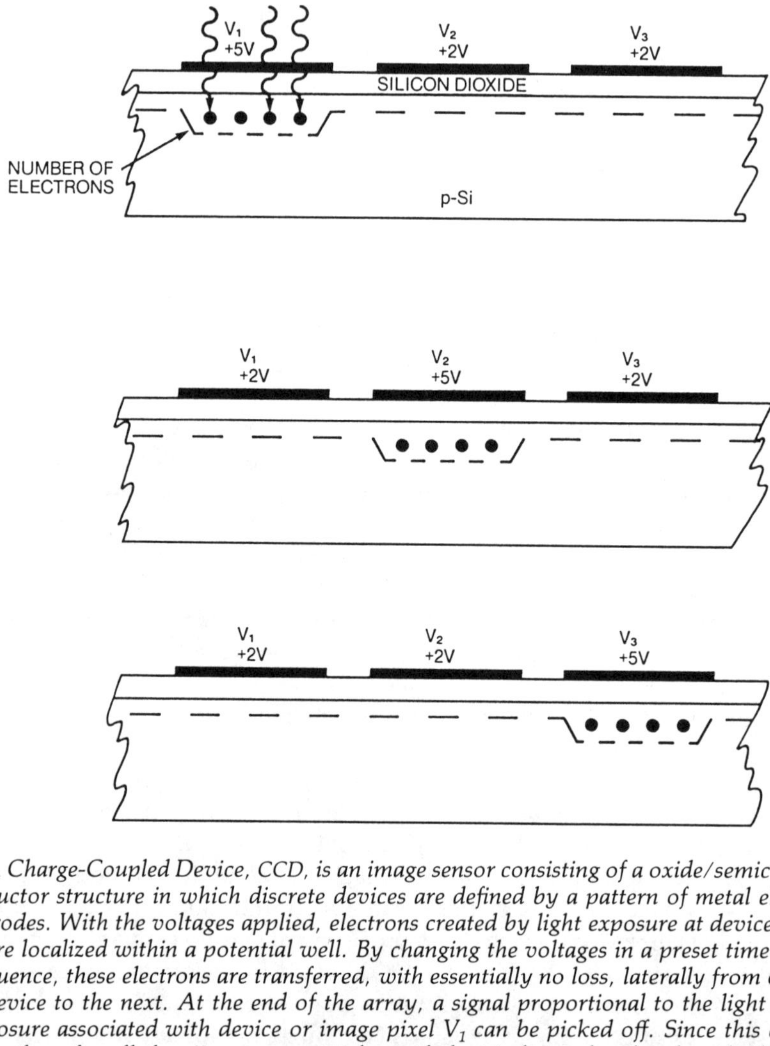

A Charge-Coupled Device, CCD, is an image sensor consisting of a oxide/semiconductor structure in which discrete devices are defined by a pattern of metal electrodes. With the voltages applied, electrons created by light exposure at device V_1 are localized within a potential well. By changing the voltages in a preset time sequence, these electrons are transferred, with essentially no loss, laterally from one device to the next. At the end of the array, a signal proportional to the light exposure associated with device or image pixel V_1 can be picked off. Since this can be done for all the sites, a sequential set of electrical signals related to the light reflected from a one-pixel wide part of a document is obtained. In this way, the total document information can be captured by passing the document beneath the CCD scanner.

on a silicon chip defined by two closely spaced gate electrodes. Light creates free carriers within a particular storage site which defines a pixel. By applying appropriate voltages to adjacent electrodes, this charge and its light-related information can be laterally transferred from one storage site to another. Since the number of carriers created at a pixel is proportional to the incident light intensity, this information can be read out sequentially and therefore provides an intensity and position dependent video signal that can be used to construct a bit map of the original image. A good analogy is that of buckets of charge being passed along the array to the end of the row, where the number of charges per pixel are measured. This is essentially a shift register, sometimes more colloquially known as a "bucket brigade." The transfer is so efficient that it can occur over several thousand sites without any loss of the light-produced carriers.

One problem with crystalline silicon-based devices is that the integrated circuit industry is more or less standardized on chip sizes about ½" square. This is not an inherent limitation, since single chip arrays up to about 2 inches square certainly can be produced, but only at a cost. As a consequence of this, 1:1 full-width arrays, up to a page-width approximately 8", are not commercially available. Letter quality reproduction of text requires a resolution of 300 pixels/inches, whereas offset quality demands about 1,200 pixels/inch. Since the original standard-size document would have to be optically focussed down to the ½" × ½" dimensions of a single chip, then the required density of active elements on the chip would have to be increased by this reduction factor of about 20. Thus the density of a single chip would need to be about 6,000 pixels per inch for letter quality, or about 24,000 pixels per inch for offset quality. Currently this is even beyond the capability of the solid-state electronics industry, quite apart from yield, reliability or cost factors. The scanning of a document therefore requires either a one-dimensional linear array with the dimensions of a page, comprised of individual chips joined or butted together, or one or more chips butted together with an overall size less than a page width.[177,181] In the former case, the scanner bar can remain stationary and the document moved under it. In the second case, the scanner may also remain stationary if image reduction is achieved using a lens. For example, a stationary one-inch linear array with a pixel density of 3600 per inch could, using reduction optics of 9:1, scan a standard page with an effective resolution of 400 pixels per inch. An advantage of detector arrays such as solid-state scanners is that the individual elements on a single silicon chip inevitably maintain an absolute position relative to each other. This reduces possible image degradation associated with errors in pixel location which is not the case in other scanning techniques such as where the original document is scanned pixel by pixel by a flying-spot scanner whose reflection is monitored by a discrete photosensitive detector.

It is the development of solid-state detectors with high resolution, compactness and low cost, coupled with the emergence of the sophisticated, high speed data processing of computers, that has enabled high-speed scanning to be married with both xerographic printing and electronic processing.[181] The coupling of an input scanner with a xerographic printer allows a printer to function, if needed, in a copier mode and the

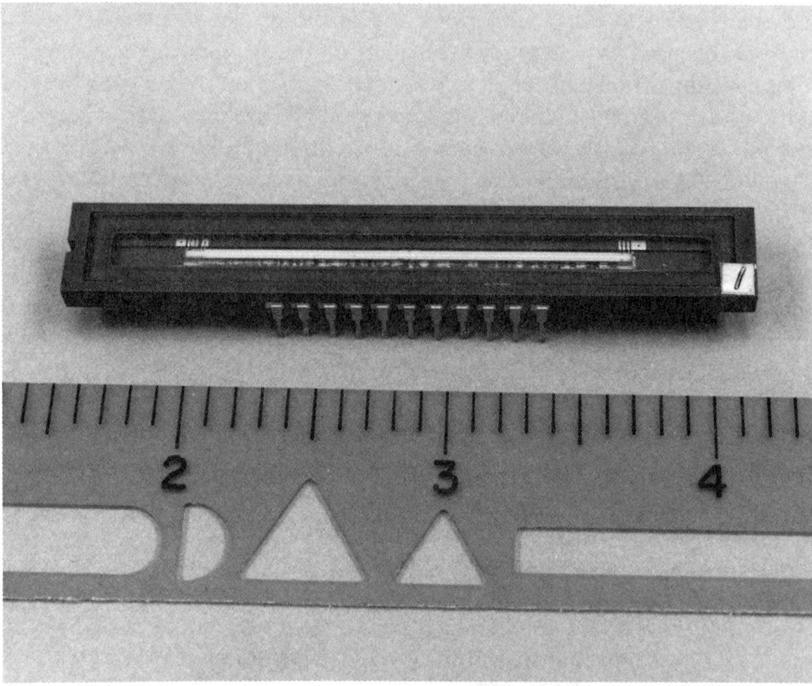

A typical solid-state scanner built on a silicon single crystal. The device contains several thousand discrete charge-coupled devices each only a few square microns in size. Since the active length is only about one and a half inches in length, reduction optics must be used to image a page-size document onto it and achieve resolutions of about 300 pixels per inch. It is also possible to butt or join several such elements to produce a full page-width scanner. In either case, a document is read by moving the scanner over it; alternatively the scanner can be stationary and the document moved.

availability of such multi-function products is growing rapidly. More sophisticated products, known as intelligent copiers, are making possible the needed link between existing documents and the world of electronic processing. The defining characteristic of an intelligent copier lies in the function of the scanner. In simple multi-function machines, the scanner is used as a passive optical-digital transducer. In the case of modern FAX machines, such digitization also enables increased speed and accuracy of transmission of original documents to remote sites for reproduction. The scanner in an intelligent copier, on the other hand, is a much more powerful device. It not only captures all the document information as a large number of pixels, but also has the capability to recognize and identify subsets of pixel combinations as characters. Intelligent copiers therefore use Optical Character Recognition or OCR scanners for documents which contain both

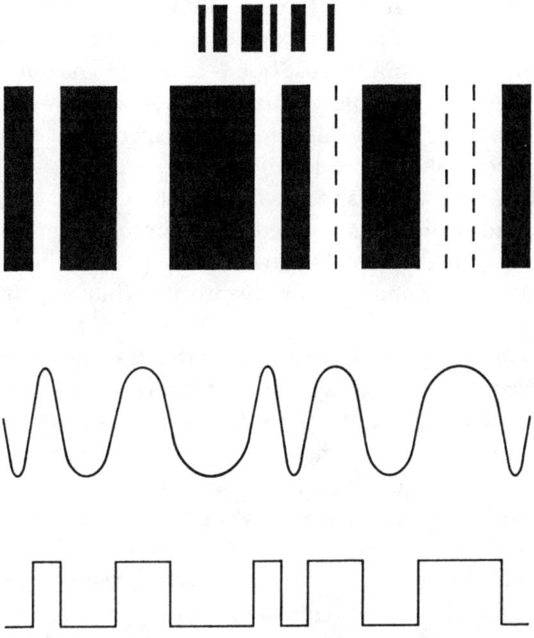

The most familiar use of optical recognition is the bar-code scanner used in stores and supermarkets. The familiar sequence of black bars and white spaces are scanned using a helium-neon laser and the reflected laser light read by a photodetector. The sequence of bars and spaces are thus transformed into the alternating voltage signal shown. This in turn can be represented as rectangular voltage signals of varying length which correlate with the specific geometric arrangements of the original bars on an item. By electronically comparing these signals with the relevant bar-code information stored in a centralized computer, appropriate merchandise information is supplied for display or print-out.

alphanumerics and line graphics while pictorial images require image scanning.

The relative ease with which we all learn as children to read belies the fact that character recognition is an immensely complex process.[182] The combination of the human eye and brain is a highly sophisticated tool for recognizing, for example, the letter "A," whether handwritten, in any of a hundred fonts or displayed at arbitrary orientations. We even have the effectively instantaneous capability to identify an illegible character, word or even a phrase based purely on its context or relationship to previous or succeeding words. OCRs are in effect replacements for this human facility and indeed the early pioneering efforts in the development of this technology, dating back to the mid-nineteenth century, were aimed at providing reading assistance to the visually impaired or the blind. A particular form of optical scanning recognition is already very much a part of our daily lives in the ubiquitous use of bar codes.[162] The familiar black bars and white spaces of variable width found on almost everything from groceries to library books use this principle. A specific set of such markings, corresponding to a number code entered into a computer, is attached to each item. The bar code is scanned with an optical wand which illuminates the bar code and projects its focused image onto a light sensor. Reading with accuracies as high as 99.99 percent these provide immediate and accurate computerized monitoring of inventories while improving customer service.

The present discussion, however, relates to the more difficult and complex task of scanning and recognizing the information content of existing printed and, ultimately, handwritten documents. An OCR scanner can address such an input document in one of the scanning modes just discussed. In doing so it produces an analog photosignal replicating the document image.[177,183] These signals are digitized by electronic sampling into binary-bit streams which consist of bit-matrix representations of the alphanumeric characters in the image. OCRs use one of two basic compare and match techniques. The first is called template matching and, as its name implies, it compares the scanned digitized image entered into a RAM against a standard digital bit-map or cut-out of letters, numerals or symbols permanently stored in a PROM. A variety of comparison techniques are employed to increase the speed and accuracy of the identification process.[183] Initially this involves using the smallest number of bits and then increasing the number until recognition of the character is made. If no definite identification can be made the machine signals an error with a star or question mark symbol and the character must be entered manually. With a successful match made, however, the binary-bit matrix is replaced with the appropriate ASCII code. This enables the unique feature of a digital copier since the transformation of the analog image into a digital character code permits manipulation of the original document by electronic processing using a computer or word processor programmed with the same character code. The substitution of the 50–100 bits per character in the binary matrix by the few bits of the character code significantly reduces the total number of bits to be handled, thus enhancing the processing speed. OCRs using this technique can do a very good job with high quality originals which use several fonts the OCR "knows," typically those most commonly used on typewriters. Unfortunately, and in part this is due to the growth of electronic documents, word processors have resulted in the creation of hundreds of new font styles. Moreover enlargement or reduction of fonts do not scale rigidly with the standard sizes, since there are relative changes within some characters. A more general approach is to use what is called intelligent character

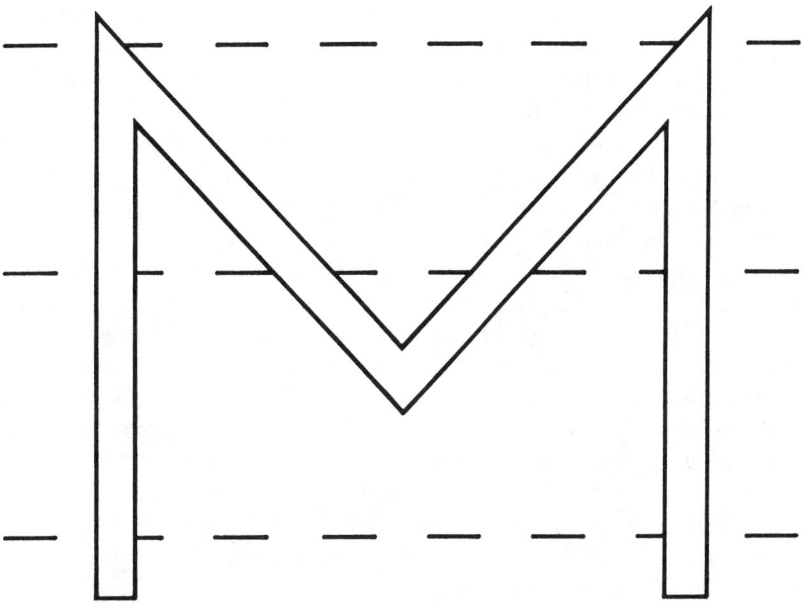

OCR or Optical Character Recognition involves the recognition of an alphanumeric symbol from a partial analysis of a character by an optical scanner. In template matching the scanned digitized image is compared with a digital bit-map or electronic cut-out permanently stored in the memory of a processing unit. A more sophisticated technique depends on the shape or topological features unique to a particular character. As shown, the letter "M" is identifiable from the combination of features encountered by scanning it horizontally, although in specific types of fonts the central minimum, for example, might have different depths.

recognition. The central distinction is that this method does not use any font-specific information. Instead, it depends on the fact that each character, letter, number or symbol has a unique set of topological features. Thus for some characters, such as the letter "A," there may be only one bit map for its ASCII code, while for others, for example the letter "M," where the central minimum may have different depths, there may be several entries in the PROM for its ASCII code. This approach therefore involves the use of more bits and can take comparatively longer than template matching but, on the other hand, permits the recognition of a much wider array of fonts.

Until now the scanning and subsequent printing has been considered in the context of text and line graphics which are largely composed of two densities, either the white of the paper or the black of characters or lines.

Such is not the case where a photograph is being scanned. The photographic process can produce a large range of gray levels depending on the amount of silver created by light in the emulsion. Faithful reproduction of a photograph therefore entails the creation of the required range of gray tones, known as the gray scale, by copying or printing technologies which have only two shades, black or white.[154] Traditionally xerographic printing has also been limited to this mode, although it is possible, in principle, by intermediate modulation of the laser beam intensity to print shades of gray.

Following the discovery of the photographic process by Daguerre in France and Talbot in England in 1839, substantial efforts were mounted to develop methods for reproducing original photographs for books and newspapers. The technique invented, called halftone screening, was developed by Talbot in about 1852 and involves something of an optical illusion.[154] Examination of any photograph in a newspaper or magazine with a magnifying glass will reveal that it is composed of many small black or colored dots. In areas which the normal unaided eye sees as black, for example, there are more black dots which appear larger, whereas in gray and white areas of the unmagnified photograph, the dots are fewer and appear smaller. The continuous tone of the photograph is simulated because, if the unmagnified dots are small enough, the eye sees the average density even though only either black or white is actually printed.

Optical halftoning for the reproduction of photographs is achieved by rephotographing the original through a screen which may be a crossline or a contact screen. The crossline screen is made of glass on which crossed opaque lines are drawn. This produces a very fine mesh of small openings or cells, just like in a screen door. The light passes through the screen before it strikes the film. Each opening in the screen acts as a small pinhole camera and the amount of light passing through determines the size of the dot produced because of diffraction and shadowing effects. The modulation is thus achieved by multiplying the optical densities of the original photograph and the screen. A high contrast film provides a threshold to produce a binary image in that above a given light intensity black dots are produced. These in turn block light when the photographic negative is used to produce the lithographic master and determine the amount of ink that will be used in any particular area. The quality of the halftone rendition is determined by the number of lines per inch. Newspaper photographs are produced with 65–85 lines per inch screens, magazine photographs with 100–120 lines per inch and high quality reproduction require 150 lines per inch and up.[144,154]

With the digital bit-maps acquired by scanning a continuous tone photograph, the halftoning is done electronically rather than optically.[184–187] The processes are analogous although there are some differences. Since the pixel size to be ultimately produced by the laser spot is fixed, the halftoning is achieved by grouping the pixels in cells to con-

Halftoning is a technique developed for reproducing the gray levels encountered in photographs when only black or white are printed. The original photograph was rephotographed using a screen consisting of criss-crossed lines. The half-tone photograph, as can be seen in the magnification, is composed of black dots whose size are largest where the original photograph was darkest; where the original was lightest the dot size is smaller. The observation of grays, therefore, is an optical illusion depending on the eye's inability to resolve the individual dots and its integration of the average density over a finite area.

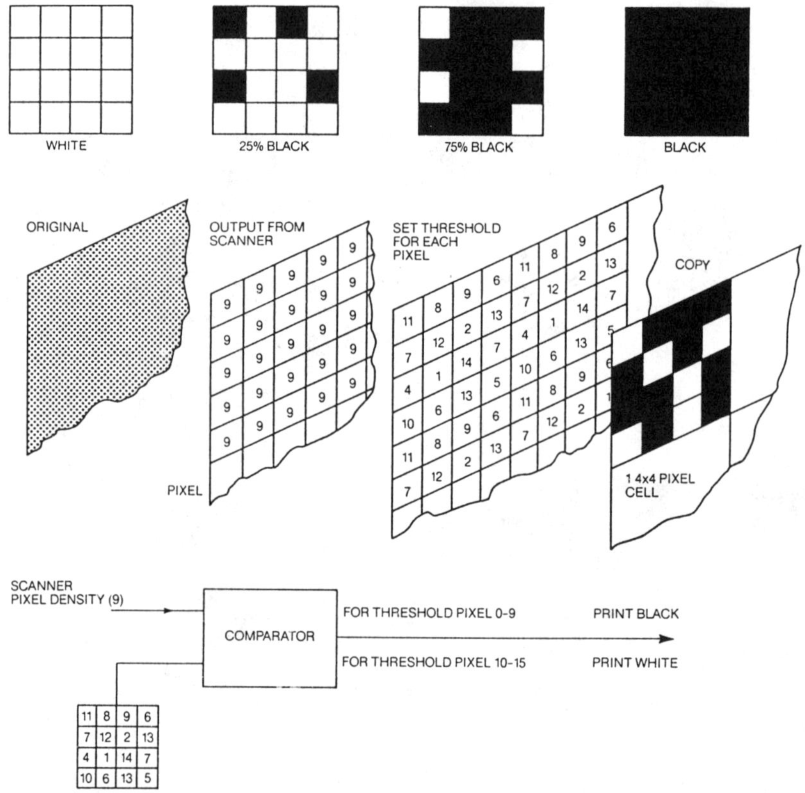

In electronic digital imaging, the halftoning can be done by electronic processing. If the analog output from a CCD scanner is digitized into a 4-bit signal, then 16 different levels of output signal ranging from 0 (white) to 15 (black) can be defined for each pixel. In order to electronically halftone, processing is carried out on a "cell" composed of a number of pixels. In the illustration, each cell contains 4×4 pixels and will appear from a distance to be some shade of gray, depending on the fraction which are black. The output from the scanner is compared with a reference or threshold grid made up of a repeating 4×4 matrix. Within a matrix, each pixel is assigned, by a random method called "dithering," a threshold level from 0 to 15. If the pixel output from the scanner equals or exceeds the level of the threshold pixel, then the output from the comparator tells the printer to print black and vice versa.

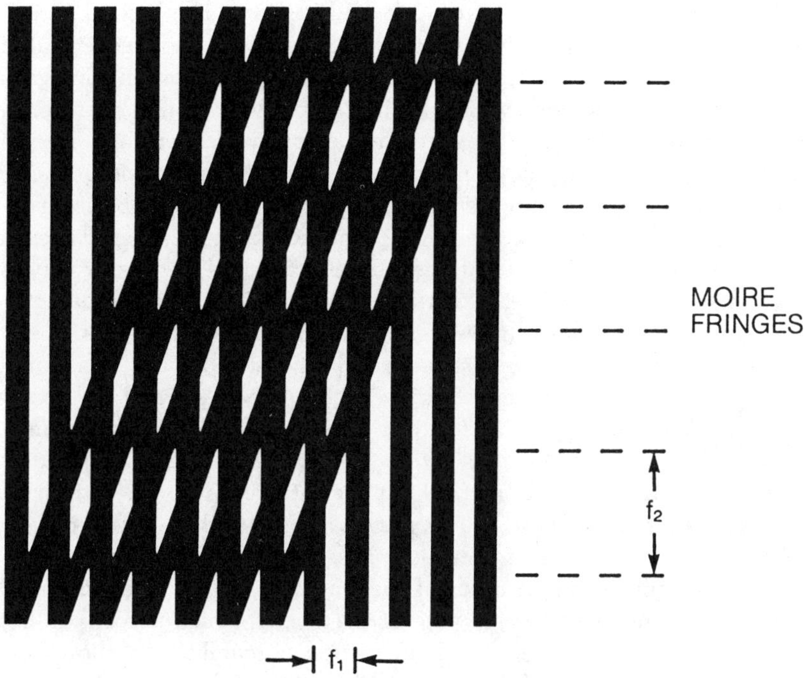

Moire fringes are observed when two screens of identical or nearly identical frequencies are crossed at an angle. If one looks at the illustration edge-on at right angle to the lines, a pattern of dark lines of is readily observable. The frequency, f_2 of these lines is lower than that of the crossed screens, f_1, and changes with the angle between them.

stitute what in effect is a screen pattern made up of such cells. This is accomplished by the use of algorithms in the scanner software. The gray levels in the scanned document determine the number of pixels within each cell which will be printed black. Thus the cells become halftone black dots of variable size and analogously to optical halftoning create the illusion of a gray scale. Digital halftoning of images can encounter two general types of problems. Although the gray scale capability increases with the number of pixels within a cell, for a fixed pixel size, so does the cell size. Under these conditions, more gray scale can only be achieved at the cost of eventual resolution loss so that the resultant image begins to look blurred. Ultimately, therefore, gray-scale printing by digital halftoning involves inevitable trade-offs between continually improving the gray-scale rendition and overall loss of image resolution and quality. The second form of generic print defect that can be encountered may be categorized as "pattern noise." For example, complicated effects can occur if the original image has information with an effective frequency content very similar to the screen frequency used. This causes gray-scale errors which visually manifest themselves as low frequency patterns modulating the printed output. These types of problems are compounded if the pictorial image to be scanned is already in the form of a halftone image. The most common everyday example of these so-called Moire patterns is the effect caused by moving very fine curtain material over itself. Work continues to design more and more sophisticated algorithms, however, to minimize these difficulties.

Except for the most sophisticated and expensive devices, OCRs cannot be used for scanning drawings, pictures, company logos and the like and these must be blanked out and entered separately with an image scanner. Clearly the advantage of OCRs is that they translate original paper documents into digital electronics without the intervention of keystrokes. A typical typist creates about 300 characters a minute while an OCR can scan and read a full letter-size page in about 15 seconds. This differential is likely to increase still further in favor of the machine. OCR scanners obviously eliminate the need for the retyping of documents which is costly in both time and resources but it also enlarges the user's capabilities by allowing the manipulation of the total document with addition, deletion or rearrangement of its contents. An important caveat is that the original document exists in an acceptable and recognizable font to the OCR scanner. Although OCR technology is developing very rapidly there are still major shortcomings. Such scanners have limited capability to recognize arbitrarily handwritten text, are highly sensitive to input data quality such as multigeneration copies of originals or wrinkled, coffee-stained documents and will respond by misreading documents either in part or whole. The number and styles of fonts are also limited as is compatibiltiy. As a result the error rates fall short of the kind of reliability one would ultimately like to have. Moreover current throughput speeds of about 5 pages/minute are restrictive but are constrained more by software issues than scanning hardware limitations.

The scanning of high resolution graphics or photographs requires prodigious amounts of memory storage capacity. The dimensions of the problems associated with scanning, storing, processing and printing electronic data can be appreciated by expressing the electronic bit information in everyday terms. A text page of the kind you are now reading contains about 40,000 bits, whereas a novelist may require 15 million bits to tell a story; a black and white photograph 500,000 bits and a newspaper magazine several hundred million. The number of bits in the Encyclopedia Brittanica (text only) has been estimated to be about 10^{11}, whilst for the

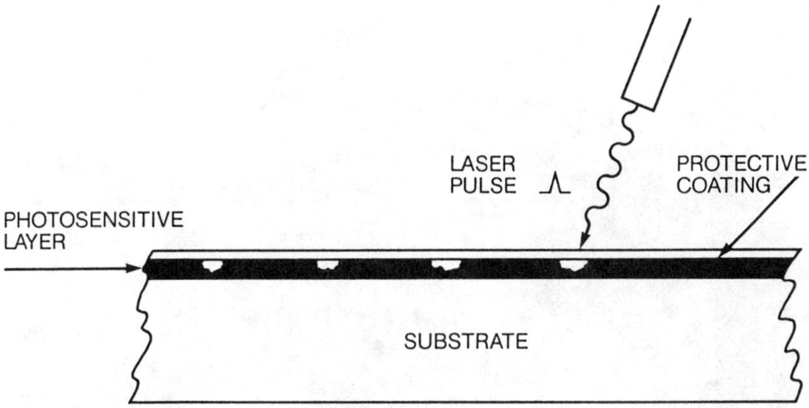

Optical disc technology capitalizes on the small size of a focused laser and its accompanying high power densities. It depends on producing localized changes in the optical properties of a thin film deposited on a disc, by writing with a laser modulated by digital information. One type involves the ablation or removal of a highly reflective material. By subsequent scanning of the used disc with a much lower power laser and the photo-detection of the reflected light, the digital information can be recovered. Given the small achievable laser spot size, enormous amounts of data can be stored on a relatively inexpensive medium. Since the photoactive layer is easily encapsulated in a tough, transparent protective layer, optical discs are virtually unaffected by surface scratches and indestructible.

National Archives the number is 10^{14} or 100 Terrabits (1 Terra$=10^{12}$). Quite apart from the computational or processing capabilities of solid-state computers, fortunately there has also been the parallel development of information storage technology. With this capability, information can not only be processed but can be stored, archivally or short term, and retrieved at will. A variety of technologies have been developed over the years such as tapes, fixed magnetic discs (including floppies), solid-state memories and more recently optical storage media.[3,4,188,189] The capacity of a memory store is defined by the number of bits it can register, although a more useful parameter is the storage density expressed as bits/sq. inch. In about two decades, the bit capacity of a MOS (metal-oxide semiconductor) RAM has increased from 256 to 256,000 while the size increase of the required chip for this thousand fold increase in capacity was negligible. In terms of storage, a floppy disc can hold about 10^7 bits whilst a hard magnetic disc can

A CD-ROM optical disc of the size shown here, 4.7 inches in diameter and 0.047 inches thick, could digitally store about 1,000 average-length novels. (Courtesy, Meckler Publishing Co., and Reference Technology Inc.,)

accommodate about 10^9. Thus ten thousand floppy discs would be needed to store the Encyclopedia Brittanica electronically.

Recent developments in memory technology have involved optical rather than magnetic media. The basic principle of optical memory storage involves the localized change in the optical properties, specifically the reflectivity, of an optical surface. This is accomplished by the modification of the physical properties of a large-area thin film by the thermal effects associated with the absorption of laser energy. Three general properties can be used.[188]

The first is ablation or pit formation in which the thin film material is literally burned away. The second involves some form of bubble formation within a thin film matrix. Finally, a phase change can be induced such as from the amorphous to crystalline phase. In all cases the two degrees of reflectivity give the binary feature of optical storage. This binary data can then read out by the same or a second laser. The discs are only 0.047 inches in thickness and the active laser layer on which the data is recorded can be overcoated by a transparent plastic coating. Since the read-out is determined by the active layer, dust, scratches and impacts do little to degrade its performance. The features of optical storage are directly related to the laser and its attributes, particularly its small size and high energy density. These allow the localized changes to be produced at relatively high writing speed over dimensions of a few square microns. As a consequence, the main advantage of these so-called optical discs is their incredibly high density information storage capabilities, so that a standard 4.7 inch diameter Compact Disc-Read Only Memory or (CD-ROM) can store 15 billion bits of data.[188,189] This is equivalent to 1,500, 5.75-inch floppy discs or 10 computer hard magnetic memory discs and almost 300,000 pages of text, the equivalent to about 1000 books of normal length, can be recorded on a single CD-ROM disc. Archival ROMs are already currently available where the information has already been recorded by the manufacturer for software storage, dictionaries and the like. Write Once, Read Many times (WORM-ROMs) which are initially blank but on which a user can record are likely to provide alternative, cheaper, more space-conserving storage than either semiconductors magnetics, or paper itself. Current optical memory storage is not without its limitations in terms of access speed and data readout rates but extensive work is ongoing to further refine and enhance optical disc technology.

Given these capabilities of scanning, optical character recognition, digital transmission, processing and storage coupled with xerographic printers, a full office automation can be envisaged.[176] The input to a word processor can come either from original key strokes on that machine or from the scanning of an existing original text or graphic input by a remote OCR/FAX systems. The word processor can be used to modify or enhance the scanned input in a variety of font styles and sizes and merge it with an original electronic document generated on the word processor. These various inputs can be merged and edited in preparation for making multiple copies using a xerographic printer. Given this combination of technologies, the emulation of publishing lies within the capability of one person as we shall discuss next.

Chapter 7
Desktop Publishing: Beyond Gutenberg

Publishing is a big business that defies any easy or comprehensive description.[190] The range of publications produced by traditional publishing is enormous and spans high quality art, specialized technical subjects, novels and children's books, in addition to newspapers and magazines. All of these are first and foremost profit-oriented enterprises. By contrast there is another class of publishing which by itself is not primarily motivated by profit. This includes the millions of reports, newsletters, catalogs and manuals produced annually by businesses, large and small, as a service to other departments and customers and which often have limited internal or public distribution. Many of these types of publications, corporate plans, price lists, schedules, insurance policies, technical drawings and the like, need to be reissued with minor or major updates on a regular basis. Such ancillary publishing activities can be a significant cost of doing business and as such certainly impact the overall profitability of an enterprise. Any process that can increase the efficiency, accuracy and timeliness with which such publications and their modifications are produced can clearly offer the possibility of substantial cost savings. These options are provided by the growth of what has come to be known as desktop publishing.

To appreciate the concept of desktop publishing, it is helpful to understand the critical elements involved in traditional publishing. Publishing is a coordinated effort involving many different tasks each of which is essential to a successful publication.[190] Usually these tasks are carried out by individuals each with a distinctive expertise. The owner or general editor of a publishing house exercises editorial control involving the identification and selection of manuscripts which meet the standards of quality, originality and marketability consistent with the mission and objectives of the firm. The general editor, as distinct from the copy editor, may acquire manuscripts from authors by a variety of means but they fall into one of two categories, solicited or unsolicited. Succinctly, the editor deals with ideas, either her or his own or the author's and ultimately decides what will or will not be published. Responsibility for coordinating the overall project falls on the general editor's shoulders who also serves as the intermediary between the publishing house and the author.

Once the final manuscript is accepted for publication, it moves to the copy editor. Here the manuscript is read carefully for spelling and punctuation errors, for areas of ambiguity in expression or development of ideas, uniformity of usage and so on. This is a balancing act in that the copy editor strives to make or suggest changes that will enhance the value to the ultimate reader, while preserving the intent and purpose of the author. Similarly the copy editor scrutinizes the projected use of illustrations and photographs for relevance, appropriateness and reproducibility. These necessarily involve some discussion and negotiation with the author.

With the completion of the copy editor's task, the modified manuscript is ready to move to the production department. Here it comes under the control of the design editor whose challenge is to fashion a book that is visually attractive as well as functionally legible. The number of words or characters in the manuscript are estimated and the type and size of font chosen to provide a readable page and a page size that will result in a book of acceptable thickness. The type of font, style of headers (chapter and page headings), quality of paper and such are considered at this point. Having decided these factors, the production department has a good estimate of the cost of production per book based on a projected press run and can establish a retail price to ensure the required profit margin. With these details settled, sample pages are set in type and printed up for approval prior to setting up and running the whole manuscript as galley proofs.

Next the manuscript is sent on to the printer who may be part of the publishing house or an externally contracted firm. The first task is composition or typesetting. This may be done using either the Linotype or Monotype process, which as their names imply involve either setting a line of type or assembling lines, one character at a time, using a punched paper tape to run a machine. Photocomposition may also be used in which images of type are made on transparent film. These are used to produce master plates for printing, usually using offset lithographic presses. If text and illustrations occur on one page then high quality proofs of the printed text are mounted and pasted together with the original artwork and this final layout is again photographed to make the final page plate. More complicated situations can arise if high quality reproduction of photographs is involved, since the necessary gravure printing involves the use of dots which does not work well with text. Accordingly a second different process may be required to reproduce text, thus increasing costs.

Photographic reproduction requires screening to produce halftone photographs. If these or line illustrations are to be published as color, then either an artist supplies the original in each of the primary colors or the final illustration is color separated by photographing the original through color filters to produce three monochrome color negatives. The final negatives of the text and pasted up text/illustrations are then used to produce the master plates.

At this point galley proofs which are samples of the text run off on long strips of paper are produced by the printer who checks them against the original edited manuscript for errors. Sets of the galley proofs are supplied to the copy editor who in turn sends a set ot the author for his proofreading. At this point "AA's" become a major concern. AA's or authors alterations can, if uncontrolled, seriously erode the profitability of a publication. Thus last minute revisions by the author at this stage can cause all kinds of complications. Even the addition or deletion of a few words can mean rearranging the type for a whole paragraph. If this occurs at the page proof stage, then the addition of a new line propagates page by page through a whole chapter. Moreover any extensive changes increase the likelihood that new errors will be introduced and because of this proofreading must be repeated. It is clear even from this very brief review that the publishing process is highly involved collection of skills and tasks which must be successfully integrated to achieve the final desired output. All the necessary tasks being labor intensive and time consuming are inherently costly. Book publishers therefore pay particular attention to the "break even point" which is the number of books which must be sold to recover the manufacturing costs, including overheads or fixed costs. Specifically most editorial and production steps are independent of the number of units produced so that the cost per book drops substantially as the run length is increased. Costs in traditional publishing are therefore highly sensitive to the projected number of units made.

The practice of traditional publishing has been systematically impacted over the last two decades by the advances in computers, lasers, and communications technologies.[3,4,144] Thus computer-controlled laser typesetting machines with printing resolutions of 1200 spi are available, although at costs ranging up to $100,000, and more and more publications such as newspapers or international magazines are published at various locations by the use of satellite communications. The distinctive requirement for desktop publishing as a viable option was the evolution of a low cost printing capability of sufficient quality. It should be stressed that desktop publishing can, in principle, be implemented, with a variety of marking technologies such as inkjet or ion deposition in addition to xerography. As laser xerography matured as a marking technology, it alone currently has the necessary combination of quality, cost and speed to enable desktop publishing systems.

Despite the fact that laser xerography originated in the seventies, desktop publishing has only begun to mushroom in the late eighties.[191–193]

Its emergence had to await the development of lower cost xerographic printers using solid-state lasers or LED arrays, more powerful and cheaper word processors, memories, scanners and associated software, together with digital transmission and networks. All of these advances were essential for the serious consideration of desktop publishing.[194] A fairly lengthy gestation period was therefore involved during which the technology was improved and extended and potential customers convinced of its virtues. In turn the customers, in an ongoing process, educated the technologists as to their needs, and the potential value they perceived in the embryonic technology. Thus an opportunity initially conceived by the technologists as a technology push, required further technical refinements to enable a truly affordable and useful product. This is one example of a technological capability which did not meet a clearly obvious or articulated need. In fact it had aspects of both technology push and market pull and typifies a more diffuse technological innovation.

An additional necessary precursor for desktop publishing involved the development and diffusion of word processors into the work place.[195] Word processors enable the production of original documents in purely electronic form. Once such a capability exists then an optical to electrical conversion step becomes unnecessary since direct communication between like word processors and an electronic printer becomes possible. Moreover, such digital electronic documents can be organized, amended or transmitted before the printing step. The widespread dissemination of such technology required the development of the word processor, a specialized form of desktop computer, to perform these tasks in an affordable, efficient and user friendly way. For these reasons, it was no accident that many of the developments in the generic area of word processors occurred not only in computer companies per se, but also in the xerographic copier industry. As early as 1973, an experimental, multipurpose work station known as the Alto was developed by the Xerox Corporation to perform many of the tasks outlined above.[196] It was designed to create and print documents containing text, images and convey such information electronically. As originally conceived, it was also intended to be used for integrated circuit design and writing programs. In other words, it was a prototype professional work station. In the present context, however, it represented one evolution of intelligent, interactive word processors requiring no special knowledge of computers or their language for operation. It was also designed to interact and communicate with like devices interconnected via a Local Area Network (LAN) consisting of a single length of coaxial cable with attached transceivers.[197]

It used a minicomputer consisting of a processor, disk drive, workstation and network transceiver. The processor memory was used to store the instructions for performing selected functions and driving printing devices.

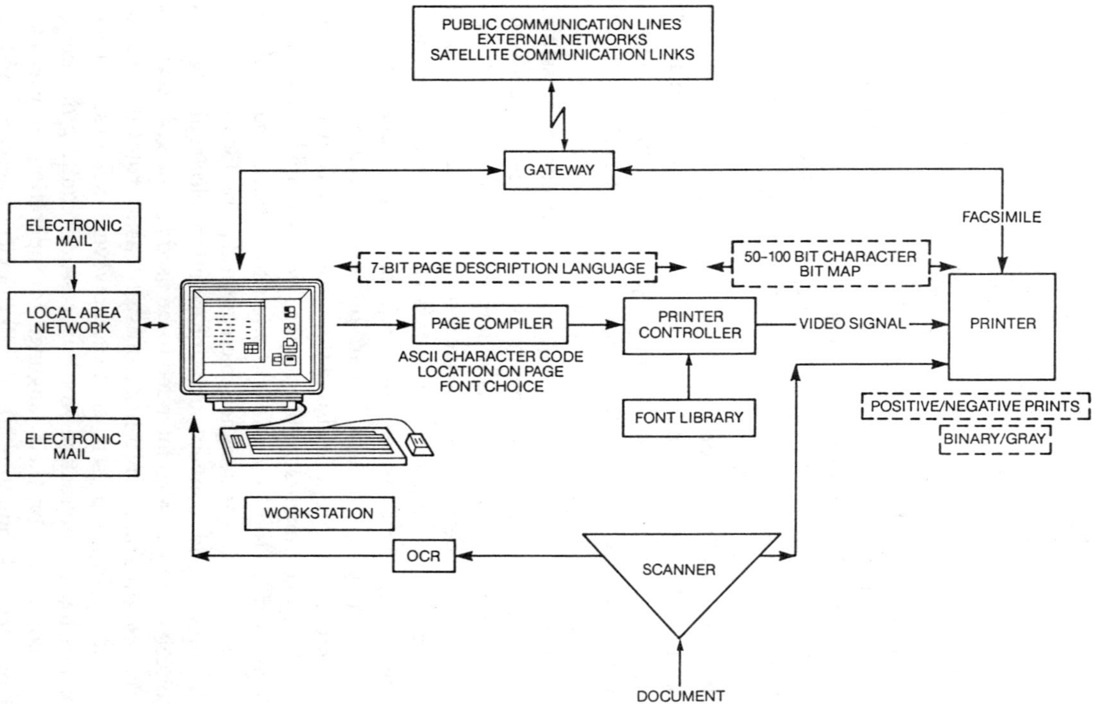

The various elements involved in document processing include a workstation which can create electronic documents as well as send or receive them from a local area or external network. By use of an OCR scanner original hard documents may also be transformed into electronic documents for any desired processing. A tie-in with a xerographic printer provides the complete integration of the tools in the hands of one individual.

The disk drive used a 2.15 megabytes magnetic storage disc from which data could be retrieved in about 100 milliseconds and transferred at a rate of 1.2 MHz. The work station included a video display refreshed at 30 times a second and a standard keyboard. The display consisted of 808 lines each of 606 individually controllable pixels to produce a 808×606 bit map. In operation it was interactive in that the user could control its function through means of what was called the "mouse." This was an electronic pointing device whose motion over a coordinate grid controlled the position of a cursor or arrow displayed on the screen. In this way the cursor could be moved over the bit-mapped CRT display to identify portions of text to which a variety of operations such as deletion, replacement, font change and various page formats could be selectively applied. The display also served to provide the operator with a true representation of the page to be printed.

Concurrently during the 1970s enormous strides were being made in solid-state electronic devices which were to make possible the necessary miniaturization and cost reduction to enable an affordable computer with the necessary computing power and yet small enough to fit on a normal desk.[195] To put this progress in perspective, a single integrated chip only 1/8" by 1/6" containing of the order of 2300 transistors can match the computing power of the first electronic computer built in the late 1940s. This contained 18,000 electronic vacuum tubes, weighed more than 30 tons and was constantly breaking down.

Even more powerful and cheaper descendants of the Alto can be connected to each other and a xerographic laser printer either directly or through a LAN called Ethernet (after the medium postulated to explain the propagation of electromagnetic waves).[197] This allows communication to other work stations, centralized computers and devices known as servers. In the event of failure, each transceiver is designed to avoid compromising the network by electronically monitoring the network to verify the status as to data transmitted or received and notify the operator. Servers include printing engines, file storage and electronic mail. The limited storage capacity on a given workstation can be augmented by using a file server. This is a centrally located storage file with an integrated capacity of 1 gigabytes or more. The individual user can store files to provide archival or intermediate storage and a capability of retrieval as needed. Privacy or security is assured by the use of passwords. With such communicating minicomputers attached to a network, electronic mail becomes possible so that memoranda and other documents can be sent to one or all stations either for direct reading on a screen or for local printout of hard copy.

By use of a word processor with instant editing, corrections, rearrangement of an electronic document and choice of fonts, the individual can wear practically all the hats involved in publishing.[176,191–193] This can

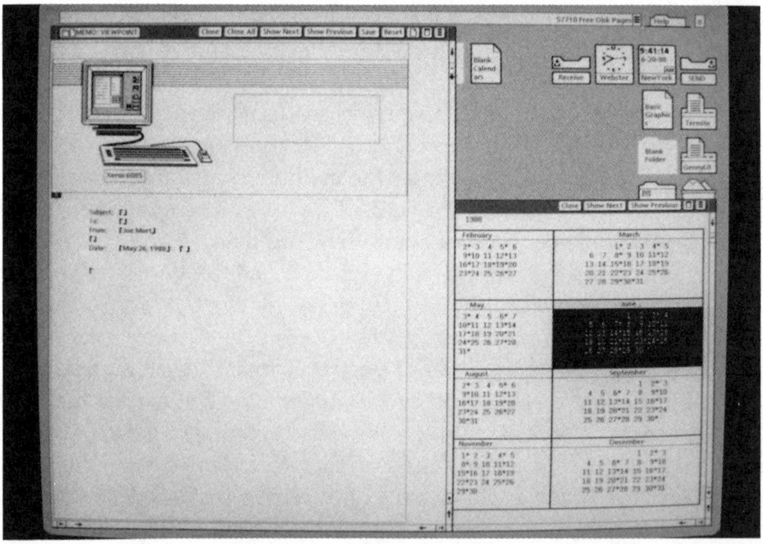

Top: *A Xerox 6500 workstation. The "mouse," visible at the right, allows the selection of icons, which visually relate to operational functions, by movement of the cursor arrow. The receive and send icons are for receiving and sending electronic mail.* Bottom: *Two windows or opened icons are also shown, one is a calendar which can be annotated by * for meetings etc., the other is a blank memorandum form.* (Photographs, Courtesy Xerox Corporation)

be accomplished on a suitably powerful word processor costing no more than few thousand dollars and small enough to place on a desk. Such word processors have evolved with user friendly configurations such that the operator does not need any familarity with computer programs but need merely point with a mouse to the appropriate icon on the screen to implement any given operation. More and more software is available to make personal computers compatible with this function. A "What You See Is What You Get" (WYSIWYG; pronounced whizzy-wig) bit-mapped screen gives real time assurance to the do-it-yourself publisher as to the desirability and acceptability of the changes being made. The power that this gives clearly short circuits the many sources of cost, delay and confusion possible in traditional publishing. Desktop publishing thus makes it possible for one individual to become author, editor, designer, production manager and distributor. In major corporations, where some publishing functions are already carried out, the people and tasks involved can be dramatically reduced with significant cuts in overheads.

It is evident that this has the potential for considerable cost saving for many publishing activities, since increasingly tasks that had to be contracted out can, for a relatively small investment, be performed in-house. Even for items that have traditionally always been done in-house such as typewritten reports and memoranda the capability now exists to significantly improve the quality and speed of production and distribution of these routine communications. Monthly sales figures, projections, inventories and budget updates can be converted into visually appealing pie or bar charts. Dramatic mixes of fonts, including bold or italics, with graphics can be employed for maximum effect. Since publishing implies the distribution of information, this can be accomplished by transmitting electronic documents, either electronic originals or those derived by scanning an original document, over a family of network-connected workstations with the option of hardcopy printing at any point. By sending inputs from several sites to a single station, individual contributions can be melded into a single document coordinated in both content and style.

Although these powerful capabilities provide the opportunity for publishing at an individual level, they also increase the burdens on the would-be publisher because of the range of skills involved. One person must become a combination general editor, creative writer, grammarian, design editor, illustrator and production department. It is a rare person indeed who possesses all of these talents naturally, so that much training will be necessary if the full benefits of desktop publishing are to be realized. For while desktop publishing clearly enables quantity, it just as surely does not guarantee quality of content. Without the recognition of such challenges, the technical facility with which documents or manuscripts may be produced has the potential of being a Damocles sword.

Chapter 8
Retrospection and the Future

Inventions, like paintings, take on a different aspect depending on how closely they are examined. From afar they appear as self-contained and clearly definable entities that spontaneously emerged from the mind and hands of one person; indeed, for almost every major invention such an individual has been identified. Closer examination, however, reveals a much more complex situation. Inventions are commonly produced in a climate of intense competition with a number of individuals striving to achieve similar ends, so that any retrospective analysis has to contend with defining the actual invention. Unfortunately the details of the environment and circumstances surrounding an invention inevitably fade with the passage of time. Yet with such understanding the technological marvels familiar to us today appear less mysterious, although no less impressive. Two examples will make the point, although it is difficult to think of an invention or scientific discovery for which a comparable story does not exist.

Gutenberg is often described as the inventor of printing. Such an attribution is inaccurate because it is imprecise. Printing, which is the exact mechanical reproduction by the transfer of designs from a single master onto paper or other medium, had its origins in the Orient and certainly predated typography.[153,154] Patterns cut in wood blocks were used to produce paper currency in China before the first century A.D. and by the Romans to make designs in plaster and on cloth. By the fifteenth century, playing cards and icons of the saints were produced from wood block cuts. Given this state of affairs, it is hard to imagine that the printing of letters was also not attempted, particularly since carving them in wood had been practiced for many years previously. In any event, the quality of such wood prints was greatly limited by the available water-based inks and smudging caused by nonuniform pressure and uncontrolled motion between the paper and blocks. The first difficulty was resolved by the use of oil-based inks, probably suggested by the use of linseed oil paints by the artists and painters of the fifteenth century. Poor quality due to nonuniform pressure was overcome by the use of the press already employed to produce textiles and paper. The problem with cutting letter sequences in wood blocks (or metals) was that each was unique since the configuration was frozen. Furthermore any mistake in production or damage in use required that the whole block be remade. By contrast the concept of movable type composed of individual letters offered the possibility of maximum flexibility for the correction of errors and reuse of the letters to produce new text. Combined with the technique of using molds to cast metallic type on demand, this approach decreased the production time and enhanced the accuracy and uniformity with which text could be created. Gutenberg's invention, therefore, was the idea of movable type, as distinct from printing or the concept of books per se. Even this attribution is at least arguable in the sense that a number of individuals were engaged in work directed towards achieving movable type and the publishing of books. Thus Johann Fust, who like Gutenberg was a goldsmith and at one time in partnership with him, is believed by some authorities to have worked independently from Gutenberg prior to their collaboration. It is also known that other related activities were carried out in the same time period, notably by Lauren Coster in Holland.

Everyone knows that Thomas Edison invented the light bulb. In fact two years before Edison's birth, in 1845, Starr in the United States had filed an American patent application describing the use of carbon filaments as light sources.[10] The first incandescent lamp was demonstrated by Staite in Britain in 1847, although it was certainly no practical device because the filament was a platinum-iridium alloy which rapidly failed because of the large amount of residual air in the bulb. This resulted from the limitations of the vacuum pumps available at that time; a problem only overcome twenty years later. In any case the required widespread distributionof electrical power from central generating stations also lay fifty years into the future. Given these shortcomings the incompatibility of expensive platinum-iridium filaments with any affordable light bulb, which was the issue addressed by carbon filaments, seems beside the point. These were two main protagonists in the commercialization of the light bulb, Thomas Edison in the United States and Joseph Swan in England. Both pursued the path of producing carbon filaments with the necessary life and cost to make incandescent lighting a practical and commercial reality and were initially embroiled in legal battles over patent coverage. These were actually resolved by an agreement to cooperate rather than fight and the Edison and Swan United Electric Light Company was established in 1883 and monopolized the nascent electric lighting industry. Even then, the final significant advance in light bulbs, the identification and use of tungsten rather than carbon filaments, was not to occur for another thirty years.[10] Parenthetically, it was during his investigations of incandescent light bulbs that Edison observed the Edison Effect which he neither understood nor pursued. This involved the observation of a current flow between the white-hot filament and a metal wire sealed into the bulb and constituted the first recorded detection of the flow of electrons thermally excited out of a metal. This occurred in 1883, 14 years before Thomson's epoch-making discovery of the electron. As the underlying principle of all vacuum tubes, it was subsequently developed by de Forest to open up the age of electronics.

The invention of xerography by Chester Carlson fits the classic mold. The use of electrostatic charging and triboelectrically charged powders to develop latent electrostatic images were certainly known and practiced prior to Carlson's emergence on the scene.[45] Carlson himself was quite clear in his own mind that it was the work of Selenyi in Hungary on the powder development of electrostatic images that gave him the idea to pursue the use of electrostatics as a better way to make copies.[44] Equally, the phenomenon of photoconductivity was recognized and well-studied, if imperfectly understood, by the 1930s. This was the era of rapid growth of interest in the photoelectronic properties of solids in both scientific and technological terms.[1,34,41,42] Just as electrical conduction in gas discharges provided a key to the unlocking of atomic structure, so photoconductivity in solids was to play an important role in confirming the energy band picture which underlies all solid-state phenomena. At the same time, the use of such photoeffects, first in vacuum cells but later in the solid-state, led to the development of devices such as the vidicon or television pickup tube. In 1933 a book was published by McGraw-Hill, entitled *Photoelectric Phenomena* by Hughes and DuBridge, which summarized the accumulated knowledge to that time.[23] This became something of technical bible for Carlson and a source of much information. As a result Carlson's invention was his perception that the combination of photoconductivity and electrostatics would be the most sensitive way of replicating projected optical images with electrostatic charges and thus provide a means of automatically making cheap copies on plain paper. With respect to the potential sensitivity of the imaging process, Carlson was unusually prescient. Counting only the energy associated with the absorbed photons, one can calculate that an energy of roughly 10^{-11} kilowatt-hours is needed to produce the electrostatic image in an average sheet of text with current, highly efficient photoreceptors. Thus the total 3 trillion latent electrostatic images currently produced world-wide in one year involves a photon energy use of about 30 kilowatt-hours. At an average utility cost of 10¢ per kilowatt-hour, the photon energy cost amounts to only about $3 for producing the electrostatic, but undeveloped, images.

In one important respect, the book by Hughes and DuBridge steered Carlson in totally the wrong direction. It explicitly states that glassy or amorphous selenium did not exhibit photoconductivity. There is, moreover, reference to the fact that the conductivity of sulfur single crystals could be increased a millionfold by illumination. Presumably influenced by these reports, Carlson concluded that sulfur was the most promising photoreceptor material. It is something of a mystery as to why amorphous selenium was classified as a nonphotoconductor. Initially one might ascribe this to a paucity of information available at such an early juncture in the development of photoconductivity. By 1927 though, the phenomenon had been studied in selenium for almost fifty years and over 1500 technical publications and about 140 issued patents had been produced.[23] Despite the fact that investigations were potentially complicated by the existence of three allotropes of selenium, the careful work of Gudden and Pohl had identified unambiguously the trigonal or gray and monoclinic forms of selenium as photoconductors.[64] Indeed Carlson, a careful and intense reader of the relevant technical literature, still thought of sulfur or anthracene as the main options for the photoreceptor as late as 1944 when his interaction with Battelle began. As is often the case in science and technology, breakthroughs which have been elusive for long periods of time suddenly seem to be achieved by a number of apparently independent efforts. Such was the case when it was ultimately discovered that amorphous selenium is in fact an outstanding photoconductor. Several groups independently made this discovery in the late 1940s and many publications on the topic appeared in the early 1950s. Notable amongst these groups were the laboratories at Battelle and the RCA Corporation.[55,67] One possible explanation of why it took so many years to ascertain this fact could be the manner in which the amorphous selenium was prepared. The early material was likely produced by solidification of molten selenium. This would have resulted in samples of considerable thickness which in turn would suggest that low applied electric fields were used. These conditions maximize the probability that the induced photoconductivity would be severely limited by trapping and hence insignificantly small.[39,40] By contrast the work at Battelle and RCA was specifically directed towards the development of very thin films for use in xerography or vidicons. In addition the production of such thin films, involving as it does the condensation of selenium vapors evaporated from molten selenium, results in the purification of the selenium as less volatile impurities remain in the molten source material.[55,56,69]

While the terms invention and innovation are sometimes used interchangeably, they are distinctly different. Nor is this merely a matter of semantic niceties, for a vast chasm exists between invention and innovation.[1,8,9] As we have seen, substantial advances had to be made to convert Carlson's embryonic idea into a viable, commerical reality. The invention Carlson offered to companies lacked an efficient means of charging the photoreceptor or any viable photoreceptor of acceptable photosensitivity. Furthermore, no suitable toner materials, development process or means of efficiently transferring developed images to paper existed. One of the steps required to bridge the gulf between invention and innovation is a technology development phase. Here, the feasibility of an invention is transformed into a practical and economical embodiment definable by the set of materials and process specifications required for manufacturing. This has to be followed, although sometimes paralleled, by a product definition involving the conceptualized use of the developed technology by customers willing to purchase it for its utilitarian value. The unsung heroes in the technological development of xerography were the engineers who had to solve innumberable problems constrained by costs, size and schedules identified for the final product.[54] Radically new manufacturing processes involving photoreceptor and developer materials had to be devised and scaled-up even while uncertainties as to all the factors controlling the materials properties or process variables remained unanswered or even unanticipated. With each of these steps the ante was raised with respect to the risk involved and the additional future resource investment required to ensure success. Carlson estimated that he had spent between $2,000 and $3,000 of his own money when Battelle came on the scene.[44] Battelle's investment was of the order of tens of thousands of dollars per year until funding by Haloid-Xerox began. In 1948 Haloid-Xerox spent development funds of $100,000, and by the introduction of the 914 copier the integrated expenditure amounted to tens of millions of dollars.

Invention and research are but two components of the innovation cycle and the least significant in both cost and risk. In isolation they can, however, be the most misleading since no matter how beguiling a technical capability may seem there is no guarantee that it will inevitably lead to a successful innovation. Even when proven technologies are identified, misjudgments of customer perceptions as to what constitutes a viable product can result in the introduction of products into markets which are either premature, ill-defined, or nonexistent. Some of the more cherished creeds of Madison Avenue are, with good reason, "give the customer what they want" or "the customer knows best." Such phrases, however, obscure a much more complex reality. In most commodity items strong customer preferences can be discerned and even measured. Products or services based on new technology are quite another matter. It can be extraordinarily difficult, for experts and nonexperts alike, to imagine or articulate in any accurate sense what value they might ascribe to a product based on a totally new technolgy.[1] Innovation has been called a game which may be either a game of chance involving skill, or a game of skill involving chance.[1] The history of xerography serves as one of the more dramatic examples of this fact. Not the least of Carlson's contributions was his tenacious belief in his invention and his advocacy in convincing others. Such faith, admirable and essential as it might have been, was no substitute for the expertise and capital resources of entrepreneurial businessmen, however. The role of Wilson and others at Haloid-Xerox, first as sponsors and catalysts of the development work at Battelle and eventually as active innovators, particularly in the area of marketing, stands as a classic in the industrial exploitation of technical invention.[56,196] Even so whatever the faith and risks taken by Wilson and Haloid-Xerox or the business judgment they exercised, there is no doubt that the magnitude of the commercial success xerography enjoyed came as a complete surprise. Thus the market surveys underestimated by a factor of almost one hundred the number of Xerox 914 machines that would be sold, while it is unclear how the real money-making parameter, the number of copies that would be made, could have been estimated in advance. Xerographic copiers and later hand-held electronic calculators are examples of situations in which customers did not know, indeed could not have known, what they wanted until they saw and more importantly used the product.

Xerography is a very sophisticated technology and is unique in the diverse properties of matter which play an essential role in its functioning. The technology is an integration of subsystems involving phenomena associated with isolated atoms as well as condensed atoms in the solid state. Thus the charging step requires the production of ions by the ionization of electrons from the essentially isolated atoms of a gas. Triboelectricity, first observed two millennia ago, also owes its origin to the divisibility of the atom. The delicate balance of electrostatic and adhesive forces important to the development process is strongly influenced by the dispersion forces caused by the mutual inducement of electrical dipoles associated with the structure of atoms. Photoreceptor operation depends on the microscopic mechanisms of charge photogeneration and motion in disordered solid-state materials, whether inorganic or molecular, which are qualitatively and quantitatively different from their ordered crystalline counterparts. The case of weakly interacting molecules in molecularly-doped polymers is particularly intriguing since molecular ions are directly involved in the conduction process. These are all interconnected with the chemical, mechanical and thermo-rheological properties of materials either in large areas, as in the case of photoreceptors, or in the micron-size particles of toner. Not only must all these work reliably in concert, but they have to be compatible with controllable and cost-effective manufacturing techniques. Even the substantial degree of understanding that now exists regarding these various phenomena cannot remove a sense of wonderment that it all works so well and dependably. Although it may not be a miracle, it still seems miraculous.

The invention of xerography took place in 1938 but the real innovation, which was the introduction of the Xerox 914 copier, occurred in 1959. It may be that the period of World War II should be discounted; however, it is arguable whether the development of Carlson's invention would have been materially accelerated had this cataclysmic event not intervened. One inference that might be drawn is that the innovation period in the case of xerography was unduly long because the degree of empiricism involved was not truly characteristic of high technology. For unlike the transistor, the significant scientific studies, just enumerated, in the fields of photoconductivity, electronic charge transport in amorphous materials, triboelectricity and structure and property relationships did not predate the technology. Quite the opposite was true in that it was the ultimate commercial success that fueled the need for less empiricism and more understanding of what is a highly complex, interrelated combination of materials and phenomena. Whereas the Nobel Prize for the invention of the transistor was received by Bardeen, Shockley and Brattain in 1951, the Nobel award for work on the physics of the disordered state to Sir Nevill Mott and Phillip Anderson did not occur until 1977.

In this respect a comparison between xerography and semiconductor technology is worthwhile for, as in many arenas, distance lends enchantment. Contrary to perception, the invention of the transistor did not immediately result in a highly sophisticated technology. The early days were marked by uncertainty in both the technology itself and even its ultimate use. Despite being a technology firmly rooted in a scientific base, the road from the laboratory to manufacturing was still paved with many pitfalls.[1,41,42] Although today only silicon comes to mind when semiconductors are mentioned, the original discovery and commercialization of semiconductor devices employed its isomorphous cousin, germanium. It was only in the mid-sixties that the number of manufactured silicon transistors exceeded those made from germanium. Thus for more than a decade, germanium was dominant despite its many shortcomings; and it was only with the invention of MOS devices and the recognition of the importance of its native oxide, which germanium lacked, that silicon technology took off.[137] In the meantime, the reproducibility and control of semiconductor devices were based, for quite a number of years, on ad hoc methods of manufacture which were as much art as science. During this very tentative period, the military, which was the dominant customer, played a unique role in nurturing the fledgling industry. As in the case of xerography, almost twenty years elapsed after the invention of the transistor before the planar technology, the integrated circuit and the displacement of germanium by silicon technology defined the full emergence of a mature semiconductor industry.

Timing is a critical element in successful innovation. This is important in the sense of making the necessary commitment of resources and anticipating the needs of the marketplace. Too much too soon, or too little too late are cliches that nonetheless capture the uncertainties of innovation. In the early phases of the research and development cycles, what may be called the commutative principle of management must be applied with great care, otherwise the results can be disastrous. This principle states simply that if the assignment of people to a task is increased by a factor of ten, for example, it will be accomplished in a tenth of the time. This can only be validly applied where progress is not knowledge-limited. It is often stated that the pace of technological innovation is accelerating year by year and that the Japanese in particular have a much shorter innovation cycle. Both of these contentions do not bear close scrutiny, if applied to any new, as opposed to derivative, technological innovation. While it may be true that the integrated technological content of our lives is growing rapidly, this is not equivalent to saying that the time required for an innovation based on a *new* technology is getting progressively shorter. In fact there is reason to believe that this is not the case because of inherent and unavoidable constraints that have little to do with technology. It is certainly possible to pursue technology development more intelligently and even to shorten parts of the innovation cycle such as the product development or engineering design stages; areas where Japan recently has had an edge. It is much more difficult to accelerate the knowledge-limited technology development stage and in most cases this can only be accomplished at the cost of introducing immature technology of unproven capability or reliability in the process, materials or ancillary subsystems. The net result is to use the marketplace itself as a development laboratory with its attendant risks.

If the time of the initial, seminal invention and the final successful innovation are carefully identified, a basic period of about 20 $\pm$ five years has remained fairly constant in the period from the start of the industrial revolution to the present. Exceptions are few and hard to find, even with the explosive growth of organized scientific and technological research since the end of World War II.[198] For the pace of scientific and technological discovery typically far exceeds the diffusion rate of technological innovation into society, as indeed it must, since society simply cannot afford to develop all the opportunities presented to it. Furthermore, market and sociological forces acting to protect sunk capital investments or head-off socially disruptive consequences are the more likely rate-limiting steps to technological change. To put this discussion in a more topical context, the discovery in 1986 of high temperature superconductors constitutes an excellent opportunity to monitor anew the time-scale for development of product innovations based on a truly new scientific and technological discovery. History would suggest that, assuming no intrinsic limitations arise with respect to the current expectations of the emergence of wires and thin films of materials with the necessary superconducting properties, the precise form of the ultimate innovation may be beyond our current imagination. It is quite likely, therefore, that many of the projected innovations so confidently predicted will not materialize. If the historical innovation period is maintained then, despite the substantial resources being poured into this field, any significant and successful product innovation based on this new technology is unlikely to occur much before the start of the 21st century.

Given these observations, the reader might reasonably conclude that there is more investment in research and technology than is needed. The story of xerography, and innovation in general, illustrate several important facts with respect to what science and technology can be realistically expected to deliver to the innovation process. Innovation is akin to a filtering process in which the pores of the filter decrease in size as the risks and costs of going further increase. The survival rate of creative or inventive ideas is, and must be, extremely low even though rejection may have little to do with technical merit. Paradoxically attempts to increase this intrinsic and therefore largely unavoidable, inefficiency can be counterproductive in that the more one tries to define the output the less likely it is to be original. Since science and technology are indispensable functions in highly technological industries, being the principle means by which a business achieves technological leadership, remains competitive or avoids technical obsolescence, this clearly puts a premium on the judgemental aspects of the management of research and technology. This is easier said than done since it is notoriously difficult for even the finest minds, whether scientist or inventor, to predict with any accuracy which areas of research are more likely to yield commercial payoff and therefore to be given priority. Thus Lord Rutherford was emphatic in his view that useful energy would never be derived from nuclear fission and Edison vehemently opposed alternating electricity on the grounds that it was too unreliable for any general system of distribution. Perhaps Michael Faraday was the most accurate when he is reported to have told Lord Palmerston that he did not know what use electrical induction might have but he was certain that the prime minister would think of a way to tax it. If chance is a factor in the game of innovation then, since luck is the residue of design, the challenge to research management is to ensure it is playing the right game, with the maximum relevant skills and as many bets placed as can be afforded. How best research management should make such choices is the subject of much debate but, although some general principles can be annunciated, no magic formula has been or is likely to be found.

Timing, in the sense of optimum commitment of resources, was alluded to earlier. In the story of xerography, as with other inventions, timing was and continues to be important from another perspective. This relates to the possible dependence of its evolution on developments in other fields. Thus the evolutionary growth of xerographic applications beyond those originally conceived by Carlson was contingent on important discoveries in science and technology which could not be foreseen. Even when the advances occurred their relevance was initially obscure, since the connections were not as obvious as they seem now. In itself the notion of using xerography in electronic publishing is less than a giant step, since Selenyi's work in the 1930s and the Stromberg-Carlson product in the late fifties clearly pointed in that direction.[46,155] Two critical elements however, were missing: the capability to effectively exploit the idea and, just as important, the need and motivation to do so. The first required a means to easily and efficiently write in raster fashion with a light beam. This had to await the discovery of the laser in 1960 and, although it was to provide the tool to acquire the capability, nothing could have been further from the thoughts of the scientists responsible.[158,160] This was followed by the necessary development phase to arrive at a practical, affordable and reliable device so that it was only in the mid-seventies that xerographic laser printers became a reality.[164,166] The emergence of a major and growing market for such printers, however, was also predicated on a growth in the use of electronic processing in the world of text and graphics. First and foremost this required the infiltration of computer technology into the office and drafting room. Again the time constant for this to occur in a substantial fashion was measured more nearly in decades than years, despite the fact there was no intrinsic technical limitation to such a penetration. The primary roadblock had more to do with costs, so that of greater importance was the rate of decline in the price of semiconductors and the increased computing power per dollar it enabled.[1] Xerographic printing and particularly its role in desktop publishing was, therefore, a consequence of the maturing of computer and xerographic technology, leading to their eventual coexistence in the office. Another important element unifying these two technologies was digital technology and its inherent high-speed data transmission capability. Semiconductor and xerographic technologies it would seem were, with hindsight, predestined to merge in the office. Perhaps an interesting harbinger of this was the fact that John Bardeen, one of the inventors of the transistor, was closely identified with xerographic technology from its earliest inception as a consultant to the Xerox Corporation and authored several patents in the area of photoreceptors.

The description of the invention and evolution of xerography and the interplay between science, technology and business given in this book was undertaken, in part, to recognize the fiftieth anniversary of the invention. It is, therefore, both natural and tempting to ask what the future might bring. Forecasting, however, is a risky enterprise at best and history records many more failures than successes. So rather than launching forth on a sea of unlimited, but predictably inaccurate, speculation it is more prudent, and perhaps ultimately more rewarding, to ruminate. Accordingly, a number of issues are discussed, none in a particularly original or profound way, which may serve to stimulate the reader to additional insights or perspectives. Certain projections which are more or less generalized extrapolations of existing or embryonic technology are relatively straightforward since one is typically dealing with the familiar. Thus continued growth in the use of xerographic products including copiers and especially printers is predictable. More and more value will accrue to the customer as increasing use of semiconductor technology is made to provide more functions that make the products increasingly user-friendly. Continuous improvement in reliability, self-diagnosis and automatic adjustment in machines to decrease the need for intervention by service personnel will also occur.

Growing needs for color copying and printing and continual advances leading to higher quality images consistent with the needs of the graphics industry will be met. Communication within local areas or extended sites will grow with the expansion of electronic information processing and transmission in the office and ultimately into the home.[199] The necessary advances in the speed and accuracy of data transmission via the increased use of microwaves, satellites and optical fiber networks can be confidently predicted.[200] Cheaper text and image scanners of enhanced capability will be developed so that widespread conversion of text and images into electronic bit streams for editing and for merging of inputs from farflung sources will allow unlimited abilities in terms of document creation.[180] All of this will necessitate continued advances in cheaper, higher density information storage with read and write capability. Digital storage technology based on optical discs appears to offer some solution to these particular problems.[188,189]

Throughout the centuries, one consistent requirement has been the need for manual selection of letters or characters, either from a printer's case or a typewriter or computer keyboard. In principle this last link with the past could be eliminated, if direct voice recognition by a machine, already demonstrated in the laboratory, was developed.[199] Without minimizing the challenges or difficulties that these advances will pose to the scientists, technologists and engineers in materials, processes, software and systems, none can be thought of as requiring other than evolutionary progress except, perhaps, the breakthroughs in artificial intelligence and related fields needed to make voice recognition a reliable, cost-effective reality.

An important component of this information age is the concept of connectivity. Connectivity is, as the name implies, the interconnection of people, sites and the multiplicity of functions enabled by digital systems and the information they create and handle. The reach of the information revolution will therefore critically depend on how rapidly such connectivity proceeds. This in turn hinges on the degree of compatibility that can be achieved given the proliferation of systems beginning to emerge. Initially, rationalization of information systems will logically occur internally within organizations. Such will almost certainly be the case for the foreseeable future since the integrated capital costs involved in assuring anything approaching complete connectivity will be astronomical. Moreover, with a rapidly developing technology the risks of making the wrong choices and ending up with obsolescent capabilities will be a further constraint on the spread of this technology. Consequently, the initial phases of this revolution will likely consist of isolated islands of internally connected information pools which will have to communicate with one another by a common medium of information exchange. Under these conditions the common currency of exchange is likely to be paper. This raises the perennial issue of a paperless society.

Given the drive to digital information and its inherent connectivity, it would seem inevitable that paper as the medium of information exchange will eventually disappear. Logically, once all information can be transmitted between desktop computers at the touch of a key for display and reading on suitable video terminals, the need for paper documents will vanish. Moreover, with the coming of a more computer literate and facile generation, the ingrained need for paper documents will naturally diminish. Logical as all this may seem, a number of arguments suggest that this may not be the scenario that unfolds. In truth, to paraphrase Mark Twain, the death of paper may be greatly exaggerated. There is a consistent tendency in technical forecasting to equate capability with inevitability and hence indulge in grossly exaggerated optimism about the form and pace of technological diffusion. If these are 'the shallow drafts that intoxicate the brain,' to quote Alexander Pope, then reality is indeed sobering. It is not difficult to assemble a list of examples to illustrate the point. One has only to reflect on the contemporary concern about the lack of collision warning systems on commercial aircraft, despite the fact that radar was invented over fifty years ago. By most estimations of the forties and fifties we should now all be driving around in electric cars and doing our grocery shopping from home via electronics. It seems incongruous for the utility company to rely on human monitoring of energy or water usage in millions of homes when there is a clear technical capability to do this automatically. The existence of the telephone for the last 100 years has not reduced the absolute amount of correspondence carried by the U.S. Postal Service. Certainly telephonic communication has reduced the volume of written material that would have been carried, but this is quite different from having completely supplanted the need for it. Anyone who has paid a visit recently to a bank or paid a bill by credit card must be struck by the continuing reliance by the clerk on paper forms while surrounded by a sea of computer terminals. Finally, in a particularly interesting paradox, digital computers are typically delivered with a stack of instruction manuals of comparable height. The list could go on and the reader will have no difficulty in identifying similar examples. Such evidence suggests that the emergence of the information revolution will not lead to an either or situation as far as paper documents are concerned.

Rather, the world of the future will be one in which paper and electronic information will and must coexist. This will certainly be true for the near term and possibly even in the long run.

In the industrial revolution of the nineteenth and twentieth centuries, machines were used to execute manufacturing processes more efficiently or, in more drastic fashion, to completely replace human functions and jobs by robots. This led to the enormous growth in industrial productivity worldwide.[7,10] One conclusion that can be drawn from a mulling over of the future is that xerography and other marking technologies, or computer technology for that matter, are not individually going to continue to be revolutionary. Rather, they will constitute parts of a revolution arising from the integration of their diverse functions into systems. Considerable evidence exists to suggest that the impact of this information revolution, already underway, may be qualitatively different from the industrial revolution.[201] More and more individuals will have, literally at their fingertips, capabilities to perform a myriad of highly complex, interrelated functions. This will democratize the production of information with potentially undreamt of speed, accuracy and efficiency. Nevertheless, despite the substantial incursion by computers and related technologies into the officeplace that has already occurred, little or no measurable increase of productivity has been discerned.[201] The explanation for this is unclear except that perhaps we are dealing with a major educational and sociological revolution with global economic ramifications cloaked in technological change.

The true revolution it seems may have to do more with how to deal with the resultant explosive growth in the creation and dissemination of information. For information processing, which is becoming just as much a reality as food processing, will profoundly affect both individual and organizational functions. Rather than replacing clearly definable and tedious capabilities, whether tightening screws or welding joints to increase manufacturing efficiency, the information revolution involves the enlargement or creation of new capabilities whose purpose and application need more precise definition. Systemic questions arise such as whether a purely information-based economy can be self-sustaining in the absence of a major manufacturing base or is it the fate of highly developed nations to become the purveyors of information, while the lower cost structures of less developed nations make them the manufacturers for the world. In any event, the nature, value and management of information and their measurement need to be better understood, for information, indispensable as it is, is not to be confused with wisdom or viewed as a substitute for judgment. At all times, attention must be paid to the customer, since the obvious but often overlooked fact is that in the end they ultimately determine whether *any* innovation is successful. More than any other market, however, the customer for information technology must be viewed by inventors, technologists, product planners, sales persons and corporate executives not merely as purchasers but active participants in its definition and application. For in the emerging world of document processing the customers are co-partners in the innovative process.

References

1. E. Braun, and S. MacDonald, *Revolution in Miniature,* Cambridge: Cambridge University Press, 1978.
2. K. Holt, H. Geschka, and G. Peterlongo, *Need Assessment,* N.Y.: John Wiley & Sons, 1984.
3. W.S. Davis, and A. McCormack, *The Information Age,* N.Y.: Addison-Wesley, 1979.
4. *Communication Technologies and Information Flow,* edited by M. Lehman, and T. J. M. Burke, New York: Pergamon, 1981.
5. P.J. Flory, "Chemistry, Macromolecules, and the Needs of Man," *Pure and Applied Chemistry* vol., 50, 255, 1978.
6. C.P. Snow, "The Two Cultures: A Second Look," in *Public Affairs,* N.Y.: Charles Scribner's, 1971.
7. J. Burke, *Connections,* Boston: Little, Brown, 1979.
8. E. Braun, "Science and Technology as Partners in Technological Innovation," *Physics and Technology,* vol., 15, 80, 1984.
9. J.A. Morton, *Organizing for Innovation,* N.Y.: McGraw-Hill, 1971.
10. T.K. Derry, and T.I. Williams, *A Short History of Technology,* Oxford: Oxford University Press, 1961.
11. I.B. Cohen, *Benjamin Franklin's Experiments,* Cambridge MA.: Harvard University Press, 1941.
12. D. Halliday, and R. Resnick, *Fundamentals of Physics,* N.Y.: John Wiley & Sons, Inc., 1981.
13. R.P. Feynman, *Feynman Lectures in Physics,* N.Y.: Addison-Wesley, 1963.
14. M. Faraday, *Experimental Researches in Electricity,* (1791—1867), N.Y.: Dover Publications Inc., 1965.
15. G. Feinberg, *Light,* Scientific American, September, p4, 1968.
16. C. Susskind, *The Early History of Electronics, II The Experiments of Hertz,* IEEE Spectrum December, 57, 1968.
17. D.A. McQuarrie, and P.A. Rock, *General Chemistry,* N.Y.: W.H. Freeman, 1984.
18. S.C. Brown, "A Short History of Gaseous Electronics" in *Gaseous Electronics,* vol. I, N.Y.: Academic Press, 1978.
19. E. Grimsehl, "A Textbook of Physics," vol. 5, *The Physics of the Atom,* London: Blackie and Son Ltd., 1935.
20. J.J. Thomson, *Carriers of Negative Electricity,* Nobel Lectures, Physics 1906 Amsterdam: Elsevier Publishing Company, 1967.
21. M. von Laue, *Concerning Detection of X-Ray Interferences,* Nobel Lectures, Physics 1914, Amsterdam: Elsevier Publishing Company, 1967.
22. M. Planck, *Genesis and State of Development of Quantum Theory,* Nobel Lectures, Physics 1918, Amsterdam: Elsevier Publishing Company, 1967.

23. A.L. Hughes, and L.A. DuBridge, *Photoelectric Phenomena*, N.Y.: McGraw-Hill Publishing Co., 1932.

24. R.M. Hochstrasser, *Behavior of Electrons in Atoms*, N.Y.: W.A. Benjamin Inc., 1964.

25. L. DeBroglie, *The Wave Nature of the Electron*, Nobel Lectures, Physics 1929, Amsterdam: Elsevier Publishing Company, 1967.

26. F.A. Wolf, *Taking the Quantum Leap: The New Physics for Non-Scientists*, N.Y.: Harper and Row, 1981.

27. J. Gribbin, *In Search of Schrödinger's Cat: Quantum Physics and Reality*, N.Y.: Bantam Books, 1984.

28. A.J. Dekker, *Solid State Physics*, London: Macmillan & Co., Ltd., 1958.

29. J.A. Crowther, *Ions, Electrons and Ionizing Radiations*, London: Edward Arnold & Co., 1952.

30. R.A. Smith, *Semiconductors*, Cambridge: Cambridge University Press, 1961.

31. W. Smith, *Effect of Light on Selenium during the Passage of an Electric Current*, Nature 7, 303, 1873.

32. W.G. Adams, and R.E. Day, "On the Action of Light on Selenium," *Proceedings of the Royal Society of London*, vol. 25, 113, 1876.

33. E.H. Putley, *The Hall Effect and Related Phenomena*, London: Butterworth Press, 1960.

34. A. Holden, *The Nature of Solids*, N.Y.: Columbia University Press, 1965.

35. W. Shockley, *Electrons and Holes in Semiconductors*, N.Y.: Van Nostrand, 1950.

36. K. Seeger, *Semiconductor Physics*, N.Y.: Springer-Verlag, 1973.

37. S.M. Sze, *Semiconductor Devices: Physics and Technology*, N.Y.: John Wiley & Sons, 1985.

38. N.F. Mott, and R.W. Gurney, *Electronic Processes in Ionic Crystals*, Oxford: Clarendon Press, 1940.

39. R.H. Bube, *Photoconductivity of Solids*, N.Y.: John Wiley & Sons, Inc., 1960.

40. A. Rose, *Concepts in Photoconductivity and Allied Problems*, N.Y.: Interscience Publishers, 1963.

41. G.L. Pearson, and W.H. Brattain, "History of Semiconductor Research," *Proceedings of the Institute of Radio Engineers*, vol. 43, 1794, 1954.

42. W. Brattain, *Discovering the Transistor . . . A Personal Account*, Datamation, June, 115, 1979.

43. A. Dinsdale, "Chester Carlson: Inventor of Xerography," *Photographic Science and Engineering*, vol. 7, 1, 1963.

44. *The Invention and Development of Xerox Copying*, Interview of Chester Carlson on December 16, 1965 by J.J. Ermenc, Copyright J.J. Ermenc, 1971.

45. C.F. Carlson, *Chapter I: History of Electrostatic Recording*, in Xerography and Related Processes, edited by J.H. Dessauer, and H.E. Clark, N.Y.: Focal Press, 1965.

46. P. Selenyi, "Application of Electrography in Television," *Wireless Engineer*, vol. 15, 303, 1938.

47. C.F. Carlson, *Electron Photography*, U.S. Patent 2,221,776, 1940.

48. C.F. Carlson, *Electrophotography*, U.S. Patent 2,297,691, 1942.

49. C.F. Carlson, *Electrophotographic Apparatus*, U.S. Patent 2,357,809, 1944.

50. R.M. Schaffert, *Electrophotography*, N.Y.: Focal Press, 1975.

51. E.M. Williams, *The Physics and Technology of Xerographic Processes*, N.Y.: Wiley-Interscience, N.Y.: John Wiley & Sons, 1984.

52. R. Hammer, "The Invention Nobody Wanted," *Fortune*, vol. 66, 155, 1962.

53. N. Langer, "Electrophotography," *Radio News*, Engineering Dept., vol. 32, 22, 1944.

54. J.H. Dessauer, *My Years with Xerox, The Billions Nobody Wanted*, N.Y.: Doubleday, 1971.

55. R.M. Schaffert, and C.D. Oughton, "Xerography: A New Principle of Photography and Graphic Reproduction," *Journal of the Optical Society of America,* vol. 38, 991, 1948.

56. W.E. Bixby, and O.A. Ullrich, Jr., *Method for the Production of a Photographic Plate,* U.S. Patent 2, 753, 278, 1956.

57. J.H. Dessauer, G.R. Mott, and H. Bogdonoff, "Xerography Today," *Photographic Engineering,* vol. 6, 250, 1955.

58. S. Lottner, *The Diffusion of the Xerox 914 Copier: A Case Study,* Rochester, N.Y.: University of Rochester Library, 1971.

59. The reader interested in the most current xerographic business and market news should consult the detailed reports issued by a number of companies such as, for example, CAP International, Marshfield, MA., Datapro Inc., Delran, N.J., Dataquest Inc., San Jose, CA., Diamond Research Corp., Oak View, CA., or Office Products Analyst, MSRA, Inc., New York, N.Y.

60. D. Adler, *Amorphous Semiconductors,* Cleveland, Ohio: Chemical Rubber Company, 1971.

61. R. Zallen, *Physics of Amorphous Solids,* N.Y.: J. Wiley & Sons, 1983.

62. H. Rawson, *Properties and Applications of Glass,* Amsterdam: Elsevier Scientific Publishing Company, 1980.

63. T.S. Moss, *Photoconductivity in the Elements,* N.Y.: Academic Press, 1952.

64. B. Gudden, and R.W. Pohl, "Uber Lichtelektrische Leitung im Selen," *Zeitschrift fur Physik,* vol. 35, 243, 1925.

65. S. Bidwell, "On the Sensitiveness of Selenium to Light, and the Development of a Similar Property in Sulphur," *Proceedings of the Physical Society,* vol. 7, 129, 1885.

66. P. Selenyi, "Photography on Selenium," *Nature,* vol. 161, 522, 1948.

67. P.K. Weimer, "Photoconductivity in Amorphous Selenium," *Physical Review,* vol. 78, 171, 1950.

68. *Photoconductivity and Related Phenomena,* edited by J. Mort and D. Pai, Amsterdam: Elsevier Publishing Co., 1975.

69. W.D. Hope, and M. Levy, "Chapter IV: Xerographic Photoreceptors Employing Selenium," in *Xerography and Related Processes,* edited by J.H. Dessauer, and H.E. Clark, N.Y.: Focal Press, 1965.

70. N.F. Mott, and E.A. Davis, *Electronic Processes in Non-Crystalline Materials,* Oxford: Oxford University Press, 1979.

71. M.A. Lampert, and P. Mark, *Current Injection in Solids,* N.Y.: Academic Press, 1970.

72. J.R. Haynes, and W. Shockley, "The Mobility and Life of Injected Holes and Electrons in Germanium," *Physical Review* vol. 81, 835, 1951.

73. R. Lawrance, and A.F. Gibson, "The Measurement of Drift Mobility in Semiconductors," *Proceedings of the Physical Society,* vol. 65B, 994, 1952.

74. J. Mort, and I. Chen, "Physics of Xerographic Photoreceptors," in *Applied Solid State Science,* vol. 5, edited by R. Wolfe, N.Y.: Academic Press, 1975.

75. W. E. Spear, "Transit Time Measurements of Charge Carriers in Amorphous Selenium Films," *Proceedings of the Physical Society,* vol. 70, 669, 1957.

76. W. E. Spear, "The Hole Mobility in Selenium," *Proceedings of the Physical Society,* vol. 76, 826, 1960.

77. D.M. Pai, and R.C. Enck, "Onsager Mechanism of Photogeneration in Amorphous Selenium," *Physical Review,* vol. B11, 5163, 1975.

78. E.J. Felty, "New Photoconductors for Xerography," *Proceedings of the 2nd International Congress on Reprography,* Cologne, edited by O. Helwich, 1967.

79. J. Mort, I. Chen, S. Grammatica, and M. Morgan, "Geminate Recombination-Controlled Photogeneration in Amorphous Solids," *Applied Physics Letters,* vol. 40, 980, 1982.

80. W.E. Spear, "Localized States in Amorphous Semiconductors," *Proceedings of*

the 5th International Conference on Amorphous and Liquid Semiconductors, Garmisch-Partenkirchen, edited by J. Stuke, and W. Brenig, vol. 1, p. 1, London: Taylor & Francis, 1974.

81. M.A. Kastner, Chapter 3: "Localized States in the Band Gap of Amorphous Semiconductors," *Noncrystalline Semiconductors,*, vol. 1, edited by M. Pollak, Boca Raton: CRC Press, 1987.

82. J. Mort and J. Knights, "Localization and Electronic Properties in Amorphous Semiconductors," *Nature* vol. 290, 659, 1981.

83. H. Hoegl, "On Photoelectric Effects in Polymers and their Sensitization by Dopants," *Journal of Physical Chemistry,* vol. 69, 755, 1965.

84. W. Mehl, and N.E. Wolff, "Photoconductivity in Dispersed Organic Systems," *Journal of Physics and Chemistry of Solids,* vol. 25, 1221, 1964.

85. R.G. Kepler, "Charge Carrier Production and Mobility in Anthracene Crystals," *Physical Review,* vol. 119, 1226, 1960.

86. W. Siebrand, "Polaron Band Structure and Carrier Mobility in Crystals of Diatomic Molecules and Aromatic Hydrocarbons," *Journal of Chemical Physics,* vol. 41, 3574, 1964.

87. *Electronic Properties of Polymers,* edited by J. Mort, and G. Pfister, Wiley-Interscience, John Wiley & Sons, Inc., 1982.

88. H. Hoegl, O. Sus, and W. Neugebauer, German Patent, 1,068,115, 1957.

89. M. Lardon, E. Lell-Doller, and J.W. Weigl, "Charge Transfer Sensitization of Some Organic Photoconductors Based on Carbazole," *Molecular Crystals,* vol. 2, 241, 1967.

90. W.D. Gill, "Drift Mobilities in the Amorphous Charge-Transfer Complex of Trinitrofluorenone and Poly(N-vinylcarbazole)," *Journal of Applied Physics,* vol. 43, 5033, 1973.

91. P.J. Regensburger, "Optical Sensitization of Charge Carrier Transport in Poly(N-vinylcarbazole),"*Photochemistry and Photobiology,* vol. 8, 429, 1968.

92. D.M. Pai, "Transient Photoconductivity in Poly(N-vinylcarbazole)," *Journal of Chemical Physics,* vol. 52, 2285, 1970.

93. J. Mort, "Transient Photoinjection of Holes from Amorphous Selenium into Poly(N-vinylcarbazole)," *Physical Review,* vol. B5, 3329, 1972.

94. J. Mort, G. Pfister, and S. Grammatica, "Charge Transport and Photogeneration in Molecularly-Doped Polymers," *Solid State Communications,* vol. 18, 693, 1976.

95. H. Seki, "Hopping Transport in Amorphous Organic Semiconductors," *Proceedings of the 5th International Conference on Amorphous and Liquid Semiconductors,* Garmisch-Partenkirchen, edited by J. Stuke, and W. Brenig, vol.2, p. 1015, London: Taylor & Francis, 1974.

96. H.Scher, and E.W. Montroll, "Anomalous Transit-Time Dispersion in Amorphous Solids," *Physical Review,* B12, 2455, 1975.

97. T. Holstein, "Studies of Polaron Motion, II, The Small Polaron," *Annals of Physics,* vol. 8, 325, N.Y., 1959.

98. W.D. Gill, "Electron Mobilities in Disordered and Crystalline Trinotrofluorenone," *Proceedings of the 5th international Conference on Amorphous and Liquid Semiconductors,* Garmisch-Partenkirchen, edited by J. Stuke and W. Brenig, vol.2, p. 901, London: Taylor & Francis, 1974.

99. J. Mort, and R.L. Emerald, "Transient Photoinjection of Electrons from Amorphous Selenium into Trinitrofluorenone," *Journal of Applied Physics,* vol. 45, 3943, 1974.

100. G. Pfister, J. Mort, and S. Grammatica, "Trap-Controlled Dispersive Hopping Transport," *Physical Review Letters,* vol. 37, 1360, 1976.

101. B.G. Bagley, "The Field-Dependent Mobility of Localized Electronic Carriers," *Solid State Communications,* vol. 8, 345, 1970.

102. S. Grammatica, and J. Mort, "Photosensitization Processes in Molecular Solids," *Journal of Chemical Physics,* vol. 67, 5628, 1977.

103. D.M. Pai, and J. Yanus, "Layered Photoconductors in Electrophotography," *Photographic Science and Engineering,* vol. 27, 14, 1983.

104. P.K. Weimer, and A.D. Cope, "Photoconductivity in Amorphous Selenium," *RCA Review,* vol. 12, 314, 1951.

105. H. Meier, *Spectral Sensitization,* N.Y.: Focal Press, 1968.

106. J. Mort, and S. Grammatica, "Infrared Sensitive Organic Photoconductor," *Applied Physics Letters,* vol. 38, 445, 1981.

107. H.F. Sterling, and R.C.G. Swan, "Chemical Vapour Deposition Promoted by r.f. Discharge," *Solid State Electronics,* vol. 8, 653, 1965.

108. R.C. Chittick, J.H. Alexander, and H.F. Sterling, "The Preparation and Properties of Amorphous Silicon," *Journal of the Electrochemical Society,* vol. 116, 77, 1969.

109. W.E. Spear, "Doped Amorphous Semiconductors," *Advances in Physics,* vol. 26, 811, 1977.

110. *Plasma Deposited Thin Films,* edited by J. Mort, and F. Jansen, Boca Raton: CRC Press, 1986.

111. W.E. Spear, and P.G. Le Comber, "Substitutional Doping of Amorphous Silicon," *Solid State Communications,* vol. 17, 1193, 1975.

112. D.E. Carlson, and C.R. Wronski, "Amorphous Silicon Solar Cell,"*Applied Physics Letters,* vol. 28, 671, 1976.

113. Y. Kuwano, "Chapter 6: Photovoltaic Structures by Plasma Deposition," in *Plasma Deposited Thin Films,* edited by J. Mort, and F. Jansen, Boca Raton: CRC Press, 1986.

114. I. Shimizu, T. Komatsu, K. Santo, and E. Inoue, "Amorphous Silicon Thin Film as a Photoreceptor for Electrophotography," *Journal of Non-Crystalline Solids,* vol. 35–36, 773, 1980.

115. J. Mort, S. Grammatica, J.C. Knights, and R. Lujan, "Studies of a-Si:H by Xerographic Discharge Techniques," *Photographic Science and Engineering,* vol.24, 241, 1980.

116. J. Mort, "Amorphous Semiconductors: From Selenium to Silicon," *The Journal of Materials Education,* vol. 5, 197, 1983.

117. J. Mort, and F. Jansen, "Chapter 7: Xerographic Applications of Plasma-Deposited Amorphous Tetrahedral Materials," in *Plasma Deposited Thin Films,* edited by J. Mort, and F. Jansen, Boca Raton: CRC Press, 1986.

118. S. Nishikawa, H. Kakinuma, T. Watanabe, K. Kaminishi, and K. Nihei, *Proceedings 34th Electronics Components Conference,* New Orleans, p. 53, 1984.

119. P.J. Flory, "Development of Concepts in Polymer Science, A Half Century in Retrospect," *Makromolecular Chemistry,* vol. 1, 5, 1986.

120. C.J. Claus, and E.F. Mayer, "Chapter XII: Liquid Development," in *Xerography and Related Processes,* edited by J.H. Dessauer, and H.E. Clark, N.Y.: Focal Press, 1965.

121. J.T. Bickmore, "Chapter XI: Aerosol Development," in *Xerography and Related Processes,* edited by J.H. Dessauer, and H.E. Clark, New York: Focal Press, 1965.

122. E.N. Wise, *Carrier Particle Development,* U.S. Patent 2,618,552, 1952.

123. C.J. Young, *Developer Brush of Magnetic Particles Agitated by Auxiliary ac. Field,* U.S. Patent, 2,786,439, 1957.

124. E. Giaimo, *Multiple, Rotating, Magnetic Developer Brushes,* U.S. Patent 2, 786,440, 1957.

125. W.R. Harper *Contact and Frictional Electrification,* Oxford: Oxford University Press, 1967.

126. J. Lowell, and A.C. Rose-Innes, "Contact Electrification," *Advances in Physics,* vol. 29, 947, 1980.

127. P.E. Shaw, "Experiments on Tribo-Electricity: The Tribo-Electric Series," *Proceedings of the Royal Society of London,* vol. A94, 16, 1917.

128. I.I. Inculet, Chapter 5: *Electrostatics and Its Applications,* edited by A.D. Moore, N.Y.: Wiley-Interscience, 1973.

129. L.B. Loeb, *Static Electricity,* Berlin: Springer-Verlag, 1958.

130. R.W. Gundlach, "Chapter IX: Development of Electrostatic Images," in *Xerography and Related Processes,* edited by J.H. Dessauer, and H.E. Clark, N.Y.: Focal Press, 1965.

131. R.B. Lewis, E.W. Connors, and R.F. Koehler, "A Spectrograph for Charge Distributions on Xerographic Toner," *Proceedings 4th International Conference on Electrophotography,* see also U.S. Patent 4,375,673, Washington D.C., 1981.

132. F.R. Ruckdeschel, and L.P. Hunter, "Contact Electrification Between Insulators: Phenomenological Aspects," *Journal of Applied Physics,* vol. 46, 4416, 1975.

133. D.A. Hays, "Electric Field Detachment of Toner," *Photographic Science and Engineering,* vol. 22, 232, 1978.

134. C.J. Mastrangelo, "Effects of Charge, Size and Shape on Toner: Photoconductor Adhesion in Electrophotographic Systems," *Photographic Science and Engineering,* vol. 26, 194, 1982.

135. J. Mahanty, and B.W. Ninham, *Dispersion Forces,* N.Y.: Academic Press, 1976.

136. G.M. Dorris, and D.G. Gray, "Adsorption, Spreading Pressure and London Force Interactions of Hydrocarbons on Cellulose and Wood Fiber Surfaces," *Journal of Colloids and Interface Science,* vol. 71, 93, 1979.

137. O.A. Ullrich, and L.E. Walkup, *A Compendium of Xerography: Development of Xerographic Images,* Columbus, Ohio: Battelle Memorial Institute, 1965.

138. M.E. Scharfe, *Electrophotography: Principles and Optimization,* Letchworth, England: Research Studies Press, 1984.

139. T. Takahashi, N. Hosomo, J. Kanbe, and T. Toymono, "Mechanism of Toner Projection Development," *Photographic Science and Engineering,* vol. 26, 5, 1982.

140. P.G. Andrus, and F.W. Hudson, "Chapter XIV: Principles of Image Transfer and Fixation," in *Xerography and Related Processes,* edited by J.H. Dessauer, and H.E. Clark, N.Y.: Focal Press, 1965.

141. J. Mammino, F. Jossel, J.M. Kobey, J.P. Serio, W.E. Solodar, and A.H. Walker, *Color Electrophotographic Imaging Process,* U.S. Patent 3,804,619, 1974.

142. J.H. Moriconi, F.M. Palermiti, and B.B. Jacknow, *Electrostatic Developer Particles Containing Resin, Colorant, Metal Salt and Phthalate,* U.S. Patent 3,609,082, 1971.

143. H.W. Hoadley, and A.S. Diamond, *Magnetographics: Study of an Emerging Technology,* Diamond Research Corporation. Ca.: Rolling Hills, 1975.

144. *Pocket Pal,* N.Y.: International Paper Company, 1974.

145. L.H. Lee, "Thermal Fixing of Electrophotographic Images," in *Adhesion Science and Technology,* edited by L.H. Lee, N.Y.: Plenum Press, 1975.

146. J.M. O'Reilly, and P.F. Erhardt, "Physical Properties of Toner Polymers," *Proceedings 2nd. International Conference on Electrophotography,* Washington D.C., p. 95, 1973.

147. R.B. Prime, "Relationships Between Toner Properties, Fuser Parameters and Fixing of Electrophotographic Images," *Photographic Science and Engineering,* vol. 27, 19, 1983.

148. D.A. Seanor, "Fusing Against Base Metal Surfaces: The Release Mechanisms," *Photographic Science and Engineering,* vol. 22, 240, 1978.

149. D.K. Donald, "Electrostatic Contribution to Powder-Particle Adhesion," *Journal of Applied Physics,* vol. 40, 3013, 1969.

150. V.M. DePalma, "An Investigation of Residual Toner Removal for the IBM 3800 Printer Subsystem," *Photographic Science and Engineering,* vol. 26, 198, 1982.

151. G. Harvapat, "A Theoretical Study of the Mechanics of a Xerographic Cleaning Blade," *IEEE-IAS Annual Conference Proceedings,* p. 569, 1977.

152. L. Nebenzahl, J. Borgioli, V.M. DePalma, K. Gong, C. Mastrangelo, and F. Pourroy, "Forces Involved in Cleaning of an Electrophotographic Layer," *Photographic Science and Engineering,* vol. 24, 293, 1980.

153. E.C. Arnold, *Ink on Paper,* N.Y.: Harper and Row, 1972.

154. W. Chappell, *A Short History of the Printed Word,* N.Y.: Alfred A. Knopf, 1970.

155. J.T. Bickmore, C.R. Mayo, G.R. Mott, and V.R. Vyerberg, "Chapter XVIII: Some Applications of Xerography," in *Xerography and Related Processes,* edited by J.H. Dessauer, and H.E. Clark, N.Y.: Focal Press, 1965.

156. R. Brown, "Lasers," *Doubleday Science Series,* N.Y.: Doubleday & Company, 1968.

157. M.J. Beesley, *Lasers and their Applications,* London: Taylor and Francis Ltd., 1976.

158. T.H. Maiman, "Stimulated Optical Radiation in Ruby," *Nature,* vol. 187, 493, 1960.

159. A. Einstein, "On the Quantum Theory of Radiation," *Zeitschrift fur Physik,* vol. 18, 121, 1917.

160. J.P. Gordon, H.J. Zeiger, and C.H. Townes, "The Maser—New Type of Microwave Amplifier," *Physical Review,* vol. 99, 1264, 1956.

161. A.L. Schawlow, and C.H. Townes, "Infrared and Optical Masers," *Physical Review,* vol. 112, 1940, 1958.

162. C.K. Harmon, and R. Adams, *Reading Between the Lines: An Introduction to Bar Code Technology,* Peterborough, N.H.: North American Technology Inc., 1984.

163. M.I. Nathan, W.P. Dumbe, G. Burns, F.H. Dill, and G. Lahser, "Stimulated Emission of Radiation from GaAs p-n Junctions," *Applied Physics Letters,* vol. 1, 62, 1962.

164. U. Vahtra, and R.F. Wolter, "Electrophotographic Process in a High Speed Printer," *IBM Journal of Research and Development,* vol. 22, 34, 1978.

165. C. Bigelow, and D. Gray, *Digital Typography,* Scientific American, August, 106, 1983.

166. G.K. Starkweather, "High-Speed Laser Printing Systems," *Laser Applications,* vol. 4, 125, 1980.

167. M. Gotlieb, C.L.M. Ireland, and J.M. Ley, *Electro-optic and Acousto-optic Scanning and Deflection,* N.Y.: Marcel Dekker Inc., 1983.

168. E.W. Williams, and R. Hall, *Luminescence and the Light Emitting Diode,* Oxford: Pergamon Press, 1978.

169. D.A. Ross, *Opto-Electronic Devices and Optical Imaging Techniques,* London: MacMillan Press Ltd., 1979.

170. W.E. Haas, "Liquid Crystal Research: The First Fifteen Years," *Molecular Crystals and Liquid Crystals,* vol. 94, 1, 1983.

171. G.M. Miller, *Handbook of Electronic Communication,* N.J.: Prentice-Hall, Inc., 1979.

172. J. Martin, *Telecommunications and the Computer,* N.J.: Prentice-Hall, Inc., 1976.

173. W.A. Triebel, and A.E. Chu, *Handbook of Semiconductor and Bubble Memories,* Englewood Cliffs, N.J.: Prentice-Hall, 1982.

174. "Printer Technologies," *Byte,* September, 161, 1987.

175. C. Meth, "Write with Light," *Administrative Management,* March, 39, 1986.

176. M. Popyk, *Up and Running,* N.Y.: Addison- Wesley, 1987.

177. D.M. Costigan, *Electronic Delivery of Documents and Graphics,* N.Y.: Van Nostrand Reinhold Company, 1978.

178. H.E. Ives, J.W. Horton, R.D. Parker, and A.B. Clark, "The Transmission of Pictures Over Telephone Lines," *Bell System Technical Journal,* April, 187, 1925.

179. D. Hunter, "Performance Characteristics of Solid-State Sensors," *Advanced Imaging*, May, p. A15, 1986.

180. *Charge-Coupled Devices and Systems*, edited by M.J. Howes, and D.V. Morgan, N.Y.: John Wiley & Sons, 1979.

181. "Scanners Take Off," *PC Magazine*, October 13, 185, 1987.

182. R.L. Gregory, *Eye and Brain*, N.Y.: McGraw-Hill, 1966.

183. A. Howard, "Character Recognition," *Systems International*, November, 69, 1986.

184. P.G. Roetling, "Binary Approximation to Continuous Tone Images," *Photographic Science and Engineering*, vol. 21, 60, 1977.

185. K.R. Castleman, *Digital Image Processing*, N.J.: Prentice-Hall Inc., 1979.

186. J.C. Stoffel, and J.F. Moreland, "A Survey of Electronic Techniques for Pictorial Image Reproduction," *IEEE Transactions on Communications*, vol. COM-29, 1898, 1981.

187. W.B. Green *Digital Imaging Processing*, N.Y.: Van Nostrand Reinhold, 1983.

188. *Essential Guide to CD-ROM*, edited by J.P. Roth, Westport, CT.: Meckler Publishing, 1986.

189. J. Hecht, "Optical Memories Vie for Data Storage," *High Technology*, August 43, 1987.

190. *What Happens in Book Publishing*, edited by C.B. Grannis, N.Y.: Columbia University Press, 1957.

191. "The Trendiness of Desktop Publishing," *The Economist*, January 24, 1987.

192. J. Seybold, and F. Dressler, *Publishing From The Desktop*, N.Y.: Bantam Books, Inc., 1987.

193. J. Felici, and E. Nace, *Desktop Publishing Skills*, N.Y.: Addison-Wesley Publishing Company, 1987.

194. C. Lu, "Laser Printers Zap the Price Barrier," *High Technology*, September, 52, 1984.

195. S. Augarten, *Bit by Bit: Illustrated History of Computers*, N.Y.: Ticknor and Fields, 1984.

196. G. Jacobson, and J. Hillkirk, *Xerox: American Samurai*, N.Y.: Macmillan and Co., Ltd., 1986.

197. *The Ethernet Handbook*, edited by N.E. Wolfberg, San Francisco: Sholwell and Associates, 1982.

198. J.J. Gilman, "Market Penetration Rates and Their Effect on Value," *Research Management*, vol. 35, 34, 1982.

199. S. Brand, *The Media Lab*, N.Y.: Viking Press, 1987.

200. M. Pyykkonen, "Networking with Light," *High Technology*, February, 25, 1986.

201. J.B. Quinn, J.J. Baruch, and P.C. Paquette, "Technology in Services," *Scientific American*, December, p. 50, 1987.

Index

A

B

C

D

G

H

I

M

N

O

P

Q

R

S

Q

R

S

T

U

V

W

X

Y

Z